NEVER MIND, WE'LL DO IT OURSELVES

NEVER MIND, WE'LL DO IT OURSELVES

The Inside Story of How a Team of Renegades Broke Rules, Shattered Barriers, and Launched a Drone Warfare Revolution

**Alec Bierbauer and
Col. Mark Cooter, USAF (Ret.)
with Michael Marks**

Skyhorse Publishing

Skyhorse Publishing books may be purchased in bulk at special discounts for sales promotion, corporate gifts, fund-raising, or educational purposes. Special editions can also be created to specifications. For details, contact the Special Sales Department, Skyhorse Publishing, 307 West 36th Street, 11th Floor, New York, NY 10018 or info@skyhorsepublishing.com.

Skyhorse® and Skyhorse Publishing® are registered trademarks of Skyhorse Publishing, Inc.®, a Delaware corporation.

Visit our website at www.skyhorsepublishing.com.
Visit the author's website at www.nevermindbook.com.

10 9 8 7 6 5 4 3

Library of Congress Cataloging-in-Publication Data is available on file.

Cover design by Brian Peterson
Cover photo credit: Greg DeSantis

Print ISBN: 978-1-5107-2091-6
Ebook ISBN: 978-1-5107-2092-3

Printed in the United States of America

To the group of dedicated Patriots who came together from all corners of the US Government to form an exceptional team and spark a technological revolution.

CONTENTS

PREFACE

We set out to write our story as a firsthand narrative, presented in alternating chapters as seen at the time through the eyes of two participants. Even shoulder to shoulder on a shared mission, an Air Force guy and an Agency guy can sometimes see the world very differently. In much the same way, the distinguished authors of our foreword, introduction, and afterword all speak about the accuracy of our story from their own unique, firsthand vantage points spanning multiple perspectives.

Charlie Allen served as the assistant director of central intelligence throughout the span of this narrative. With more than four decades of exceptional service at the CIA under fifteen different directors, Charlie knew the organization, authorities, and personalities. Additionally, as the chair of the National Intelligence Collection Board, his leadership, guidance, and blunt talk were critical to overcoming internal and external challenges. We could imagine no one better suited to comment on this book from the CIA's perspective.

Air Force Lieutenant General John "Soup" Campbell was indispensable to the evolution of the Predator drone program. He provided vision, leadership and crucial "top cover" to a team that lived on the frayed edge of the rulebook. With the unique distinction of holding senior leadership positions within the USAF and CIA,

General Campbell was both "on the field" and "watching from the box seats" at the same time. His perspective on the Predator program encompasses the strategic and the tactical.

United States Air Force Staff Sergeant Gabe Brown (now a lieutenant colonel) entered this story facing unsurmountable odds on the top of a desolate mountain in Afghanistan, standing shoulder to shoulder with a valiant team of fellow Special Operations warriors. What happened on the windswept rocks of Takur Ghar was a testament to unbreakable American heroism and tenacity. It was our great honor to help support that band of brothers in its darkest moment. If anyone can speak of the Predator story straight from the foxhole, it's Gabe.

FOREWORD

CHARLIE ALLEN

Nineteen years ago the world stood still on September 11, 2001. Islamic extremists had hijacked four United States civilian airliners, filled with hundreds of innocent Americans, and used them as weapons of mass destruction. Nearly three thousand people, mostly Americans, died. As the director of central intelligence for collection at the Central Intelligence Agency, I felt nothing but abject failure on September, 11, 2001, a failure that haunts me today, but I also felt another emotion—cold anger and the determination to destroy those who perpetrated the attack: Usama bin Laden and his inner circle of al-Qaeda supporters, the evil terrorist organization that bin Laden headed. But I have intimate knowledge of one development that was tightly held in CIA and DOD on September 11, which gave me hope in hours of despair—an ungainly aircraft called the Predator.

Never Mind, We'll Do It Ourselves is quite different from other books or articles written about the joint effort of CIA and the US Air Force to radically change operations against al-Qaeda and bin Laden. It is not a story about senior policy and intelligence officials driving decisions and making courageous decisions about the employment of the Predator to change the collection dynamic against an evil terrorist organization. Rather, it is a vivid story of midlevel men and women of the intelligence community and the Department of Defense putting aside their personal career ambitions to bring together at the working level technology, science, and true grit to create a revolution in warfighting. It is the story of how over a dozen intelligence and military organizations engaged in constant collaboration at the program management level to overcome entrenched bureaucracies and the "not here syndrome" prevalent among certain agencies. The stellar effort was unprecedented. Many of the heroes here were GS-9/11s, O-3/4s, and E-5 noncommissioned officers.

The leading protagonist in this effort was an officer named Alec Bierbauer. A former Army warrant officer associated with intelligence support to Special Operations, Alec was always like magic, appearing when you urgently needed advice and ideas on how to bring justice to al-Qaeda. When he showed up at my office at CIA headquarters in early February 2000, he seemed to know that I was responding to an urgent request from Special Assistant to the President for Counterterrorism Richard "Dick" Clarke. Clarke, hard-driving, impatient, and often short on diplomacy, had sent me a memorandum, giving me thirty days to come up with "new innovative ideas" for locating bin Laden. It was a typical Clarke memo—brusque and to the point. I told Alec that I had prepared a list of technical operations that, if implemented, would help achieve Clarke's objective. I told him that I had appointments to see the J3 and J39 in the Joint Chiefs of Staff—namely, Vice Admiral Scott Fry and Brigadier General (USAF) Scott Gration, respectively. Alec immediately launched into a long, passionate briefing on how the Predator could be used to change the collection dynamic against al-Qaeda and urged it be given top priority in my report to Clarke.

When I got to the Pentagon, I was royally received by Admiral Fry and General Gration. Both raised the topic of using the Predator against bin Laden, and the spiel was the same as that given by Alec. Why keep submarines "in the basket" off Pakistan if a few Predators could be seconded with US Air Force crews to CIA to be operated under CIA's special authorities? Gration, in particular, stated that two or three Predators, located at Indian Springs, Nevada, could be "loaned" to CIA to change the "collection dynamic." Admiral Fry stated that he was frustrated with the Agency's current lack of access and noted that Tomahawk Land Attack Missiles (TLAMs) would fly from the submarines off the coast of Pakistan if we obtained solid evidence of bin Laden's precise location; the TLAMs would destroy bin Laden and al-Qaeda's central core. I drove back to Langley knowing that "employing the Predator" would be by far atop the list of the report I would send to Clarke.

Major policy decisions were made, at times haltingly and with lawyers occasionally throwing a wrench in the works, in the spring and summer. From the first flight, I knew Alec had been right— the collection dynamic had truly changed. The United States had key al-Qaeda leaders under surveillance, and they were completely unaware of Predator's presence.

Colonel Mark Cooter is the second protagonist in this story. An Air Force major at the time, Cooter is a brash, irreverent officer (now retired) who bleeds blue but is an absolutely dedicated officer who had had significant Predator experience beginning in Bosnia. As the operations officer for the team, he was a strong advocate of arming the Predator. In early 2001, tests were conducted by the Air Force and CIA at sites in the western United States. These tests demonstrated the Predator had lethal capabilities, but explicit rules of engagement were missing. As if by magic again, Lieutenant General John Campbell, an Air Force F-15 Eagle pilot, showed up as the associate director of central intelligence for military support. He helped develop strict rules in the use of the Predator's lethal capabilities. He not only did this with aplomb but also worked to improve relations between the Air Force and CIA on all aspects of the Predator program. In addition, he recognized the value of supporting the mid-level Air Force and CIA officers who were operationally bringing the Predator project together.

In the opening chapter, entitled "First Blood," Colonel Cooter captures vividly the first day, October 7, 2001—the date the US attacked in response to 9/11 to take down the Taliban and to kill bin Laden and al-Qaeda's leaders. The Predator, guided by an Air Force pilot seven thousand miles away, was tracking the movement of Mullah Omar, the head of the Taliban, whose motorcade had stopped at unfamiliar compounds inside Kandahar. Cooter's description of the Ground Control Station and the Global Response Center is captured literally moment by moment. As Cooter puts it, "It was the first shot of a revolution that truly would be heard around the world. The face of aerial warfare had just changed forever." Not only was the Predator key to pinpointing the enemy in difficult terrain, but also it helped to guide a number of US strike aircraft into close-quarters combat, ensuring the enemy was attacked, not US SEALs and Rangers.

Never Mind, We'll Do It Ourselves should appeal to a wide set of readers, not just those involved in military, intelligence, or national security generally but also average Americans concerned by the nation's security in a world of increasing threat. Every reader should be buoyed by the intensity of this book—the absolute patriotism, dedication, and zeal of Alec, Cooter, and others. For example, the authors praise the unfailing support they received from Cofer Black, the charismatic leader of the Counterterrorism Center, and Rich Blee, a deputy, both CIA officers of the highest integrity and of operational skill. Also acknowledged by the authors is Diane Killip, a now retired officer of the Clandestine Service, whose knowledge of al-Qaeda's leaders and its network seemed unlimited.

Charlie Allen
Assistant Director of Central Intelligence,
1998–2005

INTRODUCTION

LIEUTENANT GENERAL JOHN CAMPBELL, USAF (RET.)

Despite a string of attacks attributed to al-Qaeda (AQ), dating back to the first World Trade Center bombing in 1993, the attacks on US embassies in 1998, and the attack on the USS *Cole* in 2000, AQ and Usama bin Laden (UBL) weren't on many front burners prior to 9/11. However, at senior levels in the Central Intelligence Agency and the National Security Council "the system was flashing red," and small pockets of individuals inside the highly classified world of the intelligence community were focused on what soon became the nation's number-one problem set. *Never Mind, We'll Do It Ourselves* chronicles the efforts of a group of these folks charged in early 2000 with leveraging new capabilities to gain actionable intelligence on UBL and AQ.

Their success with the small, gangly, unarmed Predator in finding and identifying one of the most sought-after individuals in the world showed the power of technology, fusion of the intelligence-collection disciplines (or "ints"), and the imagination of people who understood the capabilities of this tool. It also highlighted the hard political and military decisions presented by UBL and the al-Qaeda network in the days before 9/11.

The realization we had UBL in our sights, though, had the effect of spurring action to prepare for the next time. In retrospect, the achievement was a combination of opportunity, operational need, and leadership in key places in the Central Intelligence Agency, Air Force, and National Security Council. This dramatic demonstration of the capabilities of the Predator platform generated the development of one of the most powerful weapons on the war on terrorists.

The armed Predator was born and grew up in a highly classified world. Parts of the story have been told in various places—congressional testimony, books, and articles, and in the media—and the outline is generally well known. Most of these accounts are at the fifty-thousand-foot level—"big picture" descriptions of the system and the way it has been used as a tool to target some of the worst bad guys who would do us harm.

This manuscript is at the opposite end of the spectrum: a detailed and highly personal account from the working-level CIA and Air Force officers, technicians, engineers, and operators who designed, developed, tested, deployed, and flew the system. Starting in late 2000, with top cover and encouragement from a few fortunately placed leaders who understood the threat, these men and women tackled myriad technical and operational problems, and by September 4, 2001, were prepared to deploy and employ the system. The hurdles that remained—thorny bureaucratic and legal issues that bedeviled the program—all vanished on September 11, and the Predator soon proved its worth.

I was privileged to be assigned from 2000–2003 to the CIA as the associate director of central intelligence for military support—basically, the senior military officer at CIA, charged with coordinating the support provided by CIA to DOD, and vice versa. Since Desert Storm, both sides have seen the wisdom of providing an intermediary to bridge the culture gap that has existed since the days of "Wild Bill" Donovan. As a career fighter pilot, I was surprised to find myself in this position, but it gave me a bird's-eye view of the events described in this narrative and an appreciation for the initiative of young officers who had a mission, a vision, top cover, and a lot of running room. It confirmed my opinion that organizational affiliation matters less than mission focus, and I was proud to help when I could.

The back story of how this manuscript came to be is a tale of its own. Aside from congressional testimony, almost none of the publicly available Predator accounts ever went through a formal security review, and people who were involved in the program have been understandably reluctant to comment or contribute. The authors of this document were cognizant of their lifetime professional obligations as cleared individuals and were determined to work through the security review process. Despite the fact that the story and the technology were well known and are almost twenty years old, to say that the initial review did not go well would be an understatement. Only the determination of the authors to document their piece of history in a way that would stand the test of time—and some pro bono legal help—led to success. It's interesting that at thirty-nine months, the security review process took three times as long as the entire armed Predator development program, from the first UBL sighting in September 2000 to first employment in October 2001.

We often are fascinated by technology, but what I have come to believe is that it's all about people. Alec, Mark, Eddie, Charlie, Snake, Gunny, Ginger, Joker, John, Boom Boom, the Man With Two Brains, and the other characters who populate this story are all patriots who have kept us safe at night. They never asked for credit, but the fact that we've not had another attack on America since 9/11 is testimony to their efforts, and it's good to see them acknowledged.

Lieutenant General John Campbell, USAF (Ret.)
Associate Director of Central Intelligence
for Military Support, 2000–2003

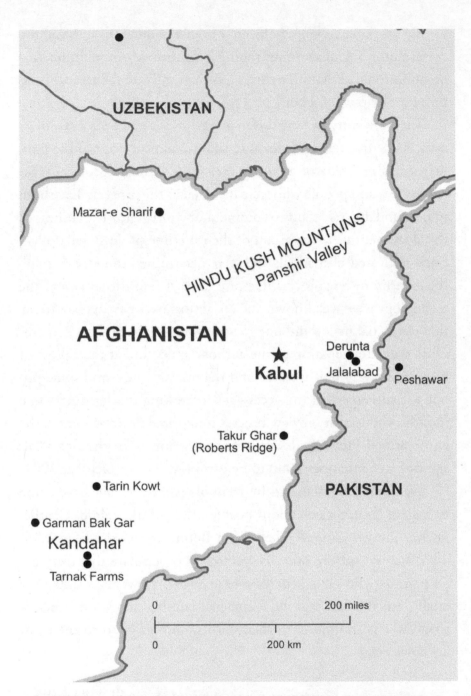

Afghanistan, noting key locations mentioned in the story. Additional information regarding the book, the authors, the characters, and the technologies can be found at http://www.nevermindbook.com.

1: FIRST BLOOD

MARK COOTER

THEN IT SHALL BE WAR
October 7, 2001

The first JDAM[1] slammed into the runway at nine in the evening Afghan time. I'll spare you the physics lesson—let's just say that when a steel pipe filled with two thousand pounds of high explosive plows into tarmac at six hundred miles an hour, you get one hell of a big hole.

In the ghostly gray world of thermal imaging, a ragged line of jet-black starbursts swept up the runways of Kandahar's military airfield. Pinpoint shots cratered taxiways and runway intersections in a single pass. The radar site that controlled the area's air and air defense units was now a smoking hole in the background. This was critical; a MiG fighter can be a dangerous opponent if it gets into the air. Our opening move was to win all the dogfights before they got started.

As an Air Force intelligence officer, I saw the video on the monitor as the money shot. The Air Force had long dreamed of real-time bomb-damage assessments, and this was as real time as you could get.

---

1   Joint Direct Attack Munition. A free-falling, all-weather "smart bomb" that can be GPS-aided and inertially guided to a specific target with great precision.

It didn't come without effort. I cajoled Alec Bierbauer, my CIA counterpart, to briefly come off our current tasking to get the shot. Our pilot, Captain Stephen "Joker" Jones, had maneuvered the aircraft to get the best angle. The SOs, sensor operators Staff Sergeant Andy R., Senior Airman Chris B., and Technical Sergeant Steve H. worked together to get the sensor into the perfect location. Captain Ginger Wallace and our operations team fed us minute-by-minute updates on the air war. That's teamwork to perfection.

The Air Force had come a long way from the days of carpet-bombing a target in the hope of hitting something vital. Unlike the dumb iron bombs of WWII that rained from the sky like bricks, JDAMs were GPS-guided, able to home in on a precise coordinate. In just over a minute, a handful of F/A-18 Hornets transformed ten thousand feet of military-grade runway into little more than a potholed stretch of old road.

That magic trick might sound like a big deal to the layman, but it was nothing new to the way America wages an air war. The JDAM had been around since 1993, when the USAF 46th Test Wing led a team of engineers at Eglin Air Force Base to bolt a guidance kit on a Rockwell GBU-15 munition. By 1997, JDAMs were in service, demonstrating a 95 percent hit ratio inside a ten-meter circle. Although it carried a price tag of less than thirty grand apiece, JDAM delivered precision targeting on par with a million-dollar TLAM cruise missile. That's a lot of bang for the buck.

What was new this night was that an Air Force major could sit in Virginia, half a world away from Afghanistan, and watch those bombs strike home. I could do so through the eyes of an airplane that had no one onboard, which was unheard of.

Less than a month had passed since the horrific attacks of 9/11, just long enough for the initial numbing shock to give way to thoughts of national security and, yeah, to exacting a measure of payback. The center point of that mission was Usama bin Laden, who was only now becoming a household name across America. But the mission to dismantle the world's most dangerous terrorist network was not limited to eradicating one man, no matter how important that guy might be. Tonight's part in that effort was focused on one Mullah Mohammed Omar.

The CIA had classified Omar as an HVT, a high value target, which in simple terms meant that he was one extremely bad character. Omar was Usama bin Laden's de facto right hand, the spiritual leader and supreme commander of the Taliban.

In Afghan terms Omar was iconic; a tall, strong mujahideen who had distinguished himself as a crack marksman with antitank rockets by destroying several Soviet armored vehicles in the 1980s. He was tough as nails, having survived numerous wounds that included the loss of his right eye to shrapnel.

Omar was also a Muslim extremist who had earlier that year demonstrated the depths of his fanaticism by attacking the Buddhas of Bamiyan, towering statues carved into a cliffside way back in the sixth century. Incredible examples of ancient Gandhara art, the Buddhas stood in silent vigil for over fourteen centuries, surviving a line of onslaughts that dated back to Genghis Khan. Sadly, that long chapter in history ended abruptly when Omar declared one of Afghanistan's oldest national treasures to be idols that offended Islam. Then he had them blown into rubble.

In an interview shortly after 9/11, Omar announced his new goal: the extinction of America. You can see why he sat atop America's hit list.

While bin Laden traveled widely, Omar rarely strayed from the familiar ground of Kandahar. And unlike the elusive bin Laden, who had seemingly vanished into thin air, we had a pretty good idea where to find Omar. Clandestine source reporting had placed him in a triangular compound called Baba Sahib, a mere thirteen miles from the airfield. The sprawling compound had been built in

1996, with some of the funding reportedly coming from the deep pockets of bin Laden himself.

Now Hollywood would have you believe that cutting-edge intelligence ops all take place in gleaming James Bond command centers lined wall-to-wall with computer screens and lots of chrome. Reality is a bit less impressive. The vital mission of ridding the world of Omar had brought my team to a lowly mobile home, complete with porta-potty and a flock of pink plastic flamingos, dumped in a parking lot in Northern Virginia. From that improbable location we were expected to drop the hammer on an enemy some seven thousand miles away.

Beyond the linear distance separating these points, the locations themselves might well have been on different planets. Virginia boasts rolling hills of lush woodlands that had already begun to shift from green to the golden hues of autumn. Kandahar, on the other hand, looked more like a bleached-out version of Mars, an endless tabletop of khaki-colored sand and dry scrub, broken only by random ridgelines of jagged gray stone.

But if the distance and disparity were not challenge enough, our mission carried one additional wrinkle. We were to fight a war not with boots on the ground but through the unblinking eye of a gangly, prop-driven, unmanned aircraft.

It is difficult to fully convey the oddity of this bird. Everything about it seemed backward, starting with the propeller back on its tail. In an age of sleek, shark-like jets, this was by all appearances the bastard child of a sailplane and a remote-controlled toy.

To appreciate its absurdity as a weapon of war, one need only compare it with a mainstay of Air Force power, the F-15 Strike Eagle. The latter has an impressive pair of turbofan jet engines that can push plane and pilot, along with over eleven tons of external fuel and ordnance, through the battle space at a blistering maximum speed of 1,900 miles per hour. In layman's terms, that's akin to a twenty-two-foot box truck, stuffed to the gills, cranking along at two and a half times the speed of sound.

By comparison, if we hung just three hundred pounds on our bird and stomped the pedal to the metal, its tiny motor would struggle to hit 90 mph. It was a John Deere tractor on a racetrack filled with Corvettes, and according to conventional wisdom one would have been hard pressed to imagine a contraption less suited for the modern battlefield.

As limited as it was in many ways, the bird that circled in the night sky over Omar's house had already evolved considerably from its even more awkward childhood in Bosnia and Kosovo. Admittedly, it was still loud as hell when you were close to it, garnering the uncharitable moniker "the flying lawnmower." And although it would never win a race to a given location, it could do things a jet couldn't dream of once on target. When a jet streaks overhead at Mach or more, the pilot is lucky to catch a brief, blurry glimpse at the ground beneath. With its ponderous speed and incredible hang time over a target, our lawnmower gave us box seats with tons of time to loiter and watch. Painted "air superiority gray" to blend in with the daytime sky, our plane was equipped with a million-dollar camera package, along with equally sophisticated communications gear that allowed us to eavesdrop on bad guys and talk to our own forces.

But the real miracle was that we were doing all this while seated in the United States, thousands of miles away from the battle space. The technology behind that breakthrough arose from a short paper written by Albert,[2] Big Safari's[3] resident mad scientist, who was reverently known in our community as "the Man with Two Brains."

---

2 We cannot disclose his true identity, so we chose Albert as a pseudonym, drawing on the next-smartest guy we could think of.

3 Big Safari is an Air Force program that provides a range of support for special-purpose weapons systems derived from existing aircraft and systems.

The final piece of evolution wasn't buried inside the aircraft but hanging underneath. Beyond the all-seeing camera pod and globe-spanning control systems, this aircraft carried Hellfire missiles.

Over most of the last two decades the laser-guided AGM-114 Hellfire had proven itself to be one of the best tank-killers in the US arsenal. The brainchild of Lockheed-Martin, Hellfire had a history of success that stretched from Panama to Iraq.

And we had a pair of them. Breaking the mold of every surveillance drone that had come before it, our bird had not come to simply watch the war but to wage it. We were about to find out if this hybrid creature could live up to its name: Predator.

I peeled my gaze off the monitor and took a moment to scan the room. The GCS[4] was alive with activity. To all outward appearance a humble twenty-foot intermodal shipping container, the GCS was crammed full of technology. The windowless steel box housed the hands-on functions of flight operations: the pilot, sensor operator, and mission commander.

In *Star Wars*, Luke Skywalker flew into battle with the ever-versatile R2-D2 in the trunk to lend a timely hand in the midst of crisis. In the real world, our lifeline was the double-wide—a mobile home parked next to the GCS, filled with the best air and air defense analysts and Afghanistan experts in the world.

Technically dubbed the Operations Cell in formal briefings, the double-wide served functions that ranged from mission planning and analysis to coordination with military elements in the battle space. To do so, it was populated with an arsenal of weather forecasters, intelligence analysts, targeters, imagery analysts, and planners. As it turned out, you need a whole lot of people to fly an unmanned aircraft.

---

4   Ground Control Station—basically, the cockpit of a plane mounted inside a steel box.

Over and around these experts, the interior of the double-wide was crammed with numerous classified computers, networked together into a veritable anaconda of bundled CAT-5 and coax cables that stretched across the ceiling. This serpent hung from a web of zip-ties and Velcro, one of many persistent reminders of the "ad hocracy" needed to achieve the impossible. Had we actually possessed a guidebook, it would certainly have included the phrase "I don't care how it looks, just make it work." I had Captain Paul Welch and Master Sergeant Cliff "Cliffy" Gross, along with contractors Pete and Martin, to thank for that.

Many of our pilots had experience flying fighters and bombers. As such, they were used to getting all their information through the comparatively narrow pipes of aircraft sensors, radios, or notes they'd compiled on a five-by-eight-foot kneeboard. They quickly became believers in the double-wide.

Alec coordinated the intelligence and political sides of the operation from within the GRC[5], a vault-like room inside CIA headquarters. Along with a slew of CIA personnel, Alec was surrounded by a team of LNOs[6] from the Air Force, the NSA,[7] USCENTCOM,[8]

---

5 Global Response Center.

6 Liaison officer, or a person who serves as a critical interface between two organizations to coordinate their activities.

7 The National Security Agency, an intelligence organization of the United States government responsible for global monitoring, collection, and processing of information and data for foreign intelligence and counterintelligence purposes.

8 The United States Central Command, a theater-level Unified Combatant Command of the US Department of Defense.

the DIA,[9] and USSOCOM,[10] along with half a dozen analysts from NIMA.[11]

These people were among the best in the world at collecting and analyzing visual and geospatial data at the speed of war. Having them all together in one room was an unprecedented act of inter-agency coordination, with "live data pipes" connecting us to some of the most important intel centers on the planet. Our housing might have been humble, but the infrastructure crammed inside was unbelievable.

My fingers drummed mechanically as a checklist of factors spooled through my mind. We had plans, back-up plans, and contingencies. If life was any teacher, not a damn one of them would be of any use exactly as written. Missions have a way of serving up the most improbable twists that force you to cannibalize parts from plans A, B, and C, then bolt them together on the fly—sometimes with a generous application of duct tape. I glanced at the clock and wondered how heavily we might have to rely on the duct tape.

My eyes tracked down, passing across the hat that hung over my console. The Stetson was new, white with a black bolo band, the kind

9   The Defense Intelligence Agency, an intelligence service of the United States specializing in defense and military intelligence.

10 The United States Special Operations Command, which is responsible for developing and employing fully capable Special Operations Forces to conduct global special operations and activities as part of the Joint Force, to support persistent, networked, and distributed Combatant Command operations and campaigns against state and nonstate actors to protect and advance US policies and objectives.

11 National Imagery and Mapping Agency, a combat-support agency under the United States Department of Defense and an intelligence agency of the United States Intelligence Community with the primary mission of collecting, analyzing, and distributing geospatial intelligence (GEOINT) in support of national security. NIMA is now known as The National Geospatial-Intelligence Agency (NGA).

of hat the Lone Ranger would wear. Much like the playoff beard to hockey fans, the hat had become something of a symbol. It had come as a birthday present from the team last year, along with a pair of silver spurs, poking fun at my cowboy demeanor. Despite numerous hoots to put them on, I refused to do so, not until we were officially in the fight. The hat was still untouched.

I glanced at my watch: 2108 local, just eight minutes since the first JDAM hit. As I stood up, I reached out and brushed my fingers across the hat.

"Ski says weather is good." Ginger's voice crackled across the intercom that connected us to the double-wide. A peerless Air Force intelligence officer, Ginger was aggregating numerous data streams pouring in from a team of experts.

I turned to face the two guys in side-by-side control seats. "Let's get on Omar's compound."

In the rightmost seat, Steve flicked his wrist, and the video feed from Predator smeared into a blur as the unblinking eye of the camera pod slewed away from the airport. Having been with the Predator program from its earliest days, Steve and I had a long history together. He had been one of my sensor-operator instructors/evaluators when I went through training, and I had deployed with him for ops in Bosnia and Kosovo. In 2000, I snagged him from a deployment to Tuzla, Bosnia, to be part of this team. With a sharp eye and a rock-steady hand, Steve was the kind of guy you wanted around when the game was on the line.

The camera view ground to a halt on Omar's retreat, a wide-angle shot that gave us a God-like view of the entire compound and much of the surrounding area. The streets were largely empty, making it easier to track a convoy of vehicles.

At first glance the view of the building seemed almost stationary, as though we had affixed a camera to a nearby cloud. This was in part due to the Predator's slow speed and in part due to the nature of the ball turret that hung down beneath its nose. The camera lens was not slaved to the direction of flight. Instead, it was free to rotate 360 degrees as the aircraft executed a precise circuit centered some distance away from the target. This would allow the aircraft to maintain a safe, and hopefully inaudible, distance while providing a persistent view of the compound.

Referred to as *the ball* for its spherical shape, the MTS[12] was a lot more than just a fancy moving camera. In addition to the electro-optical camera, the MTS provided infrared (IR) imaging, along with laser designation, rangefinder, and illumination. If you needed to "find and fix" a target, the Raytheon ball was your Swiss Army knife.

In another snap of motion, the camera zoomed in, framing the perimeter of the compound. What just a heartbeat ago had been tiny postage stamps suddenly resolved into buildings and vehicles. Still, how much detail you could resolve in the infrared spectrum was limited; most objects appeared as blobs in varying shades of grey. At the moment, we were running in "black-hot" mode, in which the hotter an object is the darker it looks on camera. It that mode the hood of a running car might appear dark grey, as would a man walking on open ground. Explosions, like the ones I'd just seen peppering the airport, looked like mid-night-black bursts of coal dust.

"I've got movement on the south side!" As a trained sensor operator with hundreds of hours' experience, Steve had the best eyes for reading the IR feed. He adjusted the camera slightly, centering on the dark grey blobs that streamed out of the compound

---

12   Multi-Spectral Targeting System, a complex set of cameras that can "see" in various spectrums, some that would otherwise be invisible to the unaided human eye. MTS includes what is commonly called "night vision," to see in extremely low-light conditions, as well as infrared thermal imaging, which can see heat in a manner similar to how people see light.

heading for a row of three parked vehicles. By a rough comparison of size, we made the center vehicle to be an SUV, likely a Land Rover, sandwiched between a pair of Toyota Hilux pickup trucks.

Through Predator's eye we saw the hoods quickly darken as engines heated up. Moments later, the three vehicles rolled out through the front gate.

"Eric, do we stay onsite or follow the convoy?" I spoke aloud to the speakerphone that kept an open line to Technical Sergeant Eric "Big E" Martin, another veteran from my previous Predator time. A towering Non-Commissioned Officer, Eric served as one of three liaison officers in the GRC. In addition to the critical roles they played, LNOs had the unenviable task of serving as a shock-absorber between Alec and me.

For several seconds the murmurs of off-mic discussion grumbled from the speakerphone before Eric came back, calm and collected as always, with "Follow the convoy."

I nodded, turning to Joker, who sat in the pilot chair just to the left of sensor ops. Joker worked the flight controls while staring intently at an array of flat-screen monitors. A former B-1 bomber pilot, Joker brought to the team a wealth of combat and tactical experience, along with an unmatched ability to coordinate with both air and ground assets. Some of the General Atomics pilots might have more time on stick with Predator, but Joker was a seasoned veteran at managing the unforgiving realities of flight in hostile airspace.

The image on the screen banked through a shallow turn, squaring up to center on the line of vehicles. The faint scratchy edge of adrenaline started to crawl up my back. The hunt was on.

An axiom in military life states that no plan survives the first contact with the enemy, and tonight proved no exception. The convoy sped away from Omar's compound, turning not out of town as expected but running straight into the heart of Kandahar.

"Where the fuck are they headed?" The words came out under my breath, to no one in particular. Then something clicked in my brain, and I turned, tossing out orders. Eric was the first in line. "Get me everything the GRC has on probable destinations—who, what, and where."

As often the case, Eric was already ahead of the question. "GRC's throwing up their hands. Too many options—the airport, a safe house; they could be going anywhere."

I nodded, understanding the realities of an urban environment, and said only, "Stay on it."

The convoy drove briskly but not in a panic, traveling just a few short miles before pulling up to another building. Dark silhouettes piled out of the vehicles and quickly disappeared into the structure. At night in IR gray, one building looked a whole lot like any of the thousand others that surrounded it.

"You gotta be fuckin' kidding me." Eric's voice garbled over the speakerphone as though he was speaking away from the mic. Then it snapped into clarity. "You're not gonna believe this. NIMA just said that the walled-in area to the left is the governor's compound."

I felt a twist in my gut. There was no chance in hell that the decision-makers were going to risk green-lighting a Hellfire shot just a hundred feet from the governor's front door. Safeguarding against collateral damage of any sort had been an unwavering concern since the beginning. Vaporizing the local governor in his own living room was way outside of our risk matrix.

To make matters worse, the building into which our bad guys just disappeared was a beast, decidedly larger than the structure we had used in our test shots. A Hellfire missile carried a roughly twenty-pound warhead. In the right configuration that's enough high explosive to burn a small hole through the side of a tank and make the interior decidedly unpleasant. Our tests back at China Lake, California, confirmed we could punch a Hellfire through the adobe-style construction common throughout Afghanistan. But given the sheer size of this place, and with no knowledge of the interior layout, we could not tell how much of the missile's lethal force would make it to people inside. Trying to engage the enemy here was shaping up to be a crapshoot on every level.

Despite the many benefits of working through an LNO, some things needed direct contact, so I told Eric to put Alec on the line. Alec came on a moment later, having barely time to say "yeah" before I hit him with a curt "What do you have on this place?"

"It's on our short list," Alec fired back in a tone that clearly revealed his multitasking. "We have credible intel that ties the site to both al-Qaeda and Taliban, but we've got bupkes on what goes on inside."

Cussing under my breath, we ran through a short conversation that ended with more "hurry up and wait." Short on alternatives and sweating the clock, we put Predator in a slow offset orbit and settled in.

You can train for a fight and execute flawlessly, but sometimes the difference between success and failure hinges on catching a break. For us that break came when the gray silhouettes suddenly, unexpectedly, scurried back to their vehicles and headed southwest out of town. We were back in business.

Tracking their movement wasn't difficult. The vehicles maintained a steady pace, showing no great haste nor making any real effort at evasion. The streets were sparsely traveled, and three vehicles moving as a group were distinctive. For a good twenty minutes they drove down the road toward Helmand Province, following a dusty stretch of pavement across table-flat dirt until they pulled into a second compound, this one much larger than the last.

What do we have here? I thought as my eyes eagerly scanned the compound. A mud wall fully enclosed the perimeter, another common Afghan feature. Inside, two rectangular single-story buildings sat in a north-south orientation. They were parallel, separated by a central corridor about the size of a soccer field. A smaller building sat at the north end of the courtyard.

The convoy pulled up to one of the two longer buildings, and once again the occupants fanned out. While picking out rifles and rocket-propelled grenades (RPGs) can be difficult in infrared, it seemed very clear that the subjects on the screen were armed. But unlike the prior stop, only a portion of the gray blobs disappeared into the building. Others, presumably security personnel, remained in or around the vehicles, weapons in hand.

Eric came up on the secure line, leading off with, "Hang on; we're working it." It was a preemptive call, Eric knowing what my first question would be. Alec's team inside CIA headquarters was glued to the same video feed we had.

I heard Alec approaching the phone, pausing to bark a question that I could make out in tone if not in words. Around him, a team of imagery analysts were stacking layers of geospatial data and reference photos in an effort to peel away the fog of war.

Alec's voice came back into clarity, and it wasn't his happy tone. "We're pretty sure the building next door is a mosque."

Figures, I growled inwardly. Terrorists were fond of hiding in places like mosques and hospitals, counting on our own paralytic fear of bad press to stay our hand. That bullshit might carry weight with the political side of the house, but to a military guy, a building is just another stack of bricks when the bad guys use it for refuge, or worse yet when they shoot at you from inside. I leaned toward the monitor, squinting at the figures pacing between the vehicles and the door.

You didn't think about our sacred places when you attacked America. But that was an argument for another place and time. Tonight wasn't about debate.

"Is that a 'we know' or a 'we think?'" I prodded Alec. Setting aside my personal opinions, the decision to rain fire from the heavens on a "culturally significant position" like a mosque or a government building was a serious proposition, one not to be taken lightly. The due diligence ahead of a green light was nonnegotiable.

"Brian[13] made the call but I think he backed his way into it," Alec said. "The phrase 'might be a minaret' popped out of his mouth, and once it did, the idea took hold. He knows it's not the popular answer, but he's sticking to his guns. We've been beating on USCENTCOM to make a decision but were told they had to wait for the JAG[14] to arrive. I can tell you up front that the book answer is no. A mosque is a clear no-go, no matter who is inside."

"Is that a 'Rich, the USCENTCOM LNO' version, or something we're stuck with?"

Alec replied curtly, "That's straight from Franks and Tenet both. A mosque is outside of ROE,[15] period."

Possessed by a possibly irrational optimism, I laid out how we could mitigate the proximity of the "mosque" by the geometry of our shot, but as the discussion with Alec dragged on it became clear that a strike on the building simply wasn't going to happen.

"All right, all right," I groused, knowing as well as Alec that the discussion was a dead-end. Special operations personnel were afforded a lot more latitude than most to decide what rules could be bent or broken, but this wasn't one of those moments—especially not with top brass watching every step. I ended the call with "Lemme know if anything changes."

---

13 Brian was a talented, young NIMA geospatial imagery expert. It took guts and conviction for "the young guy" to hold to his assessment in an atmosphere in which others were clearly looking for a different conclusion.

14 Judge Advocate General, a member of the military concerned with military justice and military law.

15 Rules of engagement, a defined set of parameters that govern when, how, to what extent, and against whom combat may be conducted.

As always in times of delay, calls from Air Force command centers started coming in. Our shield from that onslaught was my commander, Colonel Ed Boyle. I didn't envy him as he took the barrage of calls from senior officials, but I sure as hell appreciated him. Since word of Predator had leaked among the upper echelon of battle-space commanders, it had become routine for general officers and colonels to call and demand that we do something for them that ran outside our tasking. Colonel Boyle would start by professionally explaining the situation from our perspective, right up to the point that the more irate callers would start throwing rank. He would do his best to work a solution but never relented from our mission. He took the heat, gave us direction when we needed it, and always supported the members of the team.

Unease built slowly in my gut. Our momentum was beginning to slip. We needed something to stave off the growing sense of stagnation.

"All right," I announced abruptly, glancing up at the clock. "We're going to be here awhile."

We were due for a crew swap. One of the most profound aspects of unmanned aerial vehicle (UAV) flight was the ability to swap out flight crews at will. Every mission carried several pilots and sensor operators who could be rotated at regular intervals to keep clear eyes and steady hands on the controls. Like a baseball coach going into the ninth inning, I looked at my bullpen. I knew the right choice was Swanson and Guay.

Air Force Captain Scott Swanson had been with Predator through the first missile tests, so he had actual practice runs pulling the trigger on a Hellfire. At the time, he was the only military pilot on the team that had actually fired a Hellfire, so he was the obvious choice. He had Special Operations–type helicopter experience and had previously served as my Weapons and Tactics officer.

Jeff "Gunny" Guay was, well . . . he was Gunny. A crusty Air Force master sergeant, Guay had been dubbed "Gunny" as being the embodiment of what you'd imagine as a hard-charging, chain-smoking Marine noncommissioned officer (NCO). In a branch of service predominated with spit-and-polish precision, Gunny was the guy we kept in a red box that carried the stencil IN CASE OF WAR, BREAK GLASS. He was wiry and tough, and if asked a question he'd tell you exactly what he thought. Sometimes he'd tell you without being asked. But whatever he might have lacked in protocol, he made up as one of the best damn sensor operators in the service.

"Let's stay sharp," I chided, having no real need to speak the words. Everyone transitioned with equal professionalism, the shift made easier by the fact that the incoming team was in most cases parked close to the hot seats, watching the same feeds as the current operators.

We had change-over checklists, fuel-remaining checks, formal steps that had been developed over time to make sure a critical detail didn't fall between the cracks. Another of our pilots, "Big,"[16] stayed in the center seat, an ad hoc support position that typically assisted with handling radio communications with other assets in the airspace as well as weapon-launch procedures.

The incoming team had no egos, no prima donnas—just men and women at the top of their game ready to be chief cook or bottle-washer to meet shifting mission demands. *Without that spirit*, I thought quietly, *none of this shit would have worked.*

Despite the brief flurry of activity, time quickly settled back to a crawl. As the minute hand finished a full lap, it was looking like we had depleted our bag of luck. The group of armed men from

---

16 Big (unrelated to "Big E") was the moniker for a lead General Atomics test pilot who had been on our 2000 team. Besides being the test pilot, he had been in the Air Defense Artillery in the Army and now was a Navy Reservist F-18 mechanic.

the convoy had not emerged from the building. Worse yet, given the size of the structure, we would be unable to know if a handful of guys departing were the same ones that came in from Omar's compound. We had maintained absolute ownership of the vehicle, along with the guys who stood around them, but the gray rectangle in the crosshairs might have had two guys inside, or twenty. We could make nothing but a bald-faced guess as to where Omar might be inside.

Unbidden, Eric came across the speakerphone with a question from deep left field: "Hey, can we hit one of the security vehicles?"

My immediate response began with "What the fuck" and quickly went downhill from there. We hadn't done all this to shoot a fucking parked car, and what dumbass came up with—

"Sir," Eric cut in emphatically. "The JAGs at USCENTCOM are going on about positive visual control of the vehicles and the clearly armed insurgents. Some of the folks here are thinking a shot on the car might flush Omar outside."

I paused mid tirade, suddenly mulling the notion. It certainly wasn't plan A. Two hours ago it might not have made the list for plans B, C, or D. But this looked like one of those moments when we reached for the duct tape.

I turned the idea in my head. Our Predator did have two Hellfires. If one of the security cars suddenly vaporized, most anybody with common sense would scramble to bug out. A guy running to another vehicle, or running off into a field, should be pretty obvious when surrounded by half a dozen armed guards.

My eyes narrowed looking at the vehicle on the screen. The chance for a second shot aside, this would be a clearly defined engagement opportunity against confirmed combatants and a chance to prove Predator in an actual theater of war. But the practical realities of military life scratched at my irritated nerves. With the topmost brass of both the Air Force and the CIA watching and expecting great things, we were going to blow up a piece of shit Toyota.

"This is the stupidest fucking thing I've ever heard," I concluded aloud, my rant threatening to pick up where it had broken off.

In a decisive tactical sidestep, Eric cut me off with "Maybe you need to talk to Alec."

I heard the rustle of his headset hit the desk, then the somewhat distant sound of Eric's voice. "Alec, Mark has a problem with the current plan."

It was a smart play on Eric's part. "Shoot the messenger" was a daily event in the no-man's-land that stretched between the Air Force and the CIA. Our three LNOs had to spend their every waking moment stuck in that minefield, managing information, authorities, and interagency diplomacy. A good chunk of their time was spent saying no to people of higher rank and absorbing the blowback. It was a bitch of a job that demanded smarts and a thick skin.

"Take me off speakerphone." Alec had his own way of dealing with the less charming aspects of my personality. I punched a button and snatched up the handset.

Alec's voice carried what I'd come to recognize as his "don't fuck with me" tone. "There's some senior-level inertia behind thumping the lead security car. I don't need the argument, just tell me yes or no. Can we hit it?"

I shifted my weight and took a moment to pause. A second ticked by, maybe two, before I replied. "Yeah. Of course, we can hit it." I said the words as if we had ever shot a car, or anything else, in actual combat. But on paper, yeah, no problem.

"All right, Tenet seems willing to kick the anthill and see what comes running out."

Alec ran through the pros and cons, and I responded with outcomes and trade-offs. I knew Alec well enough by this point to know he was working for the best possible outcome, navigating his own political minefield. That hazard was one I very much preferred to avoid. Though I'd never admit it, I was glad to have him out there.

He came back with a terse "hang tight," and the line went silent. I returned the handset to the cradle and set the line back to speaker mode.

I took a deep breath and looked around the GCS, leaning forward with my own plans. If there was one certainty in working a kinetic engagement this high up the political and military food chain, it was that the time required for them to decide would equal, if not just exceed, our remaining fuel supply. Worse, if they came back with a green light, the need to shoot would retroactively become mission-critical despite the plane's empty gas tank. So managing fuel consumption moved even further up the priority ladder.

"Listen up, guys," I said loudly, stepping up and tapping the large screen on the wall. "We're working this security vehicle as a possible target. Let's get ahead of the call and run our AGM-114 pre-launch checklist[17] and descend to our shoot altitude."

We had been parked in a high orbit that optimized fuel savings and invisibility. In prepping to fire, we would have to descend to a lower altitude where the racket of our engine might draw attention. The guards still loitered around the target vehicle. They may be little more than blobs on the screen in front of me, but out there in the real world they were bad guys standing outside in the quiet of night.

"Gunny, keep your eyes glued to the building and shout out if anything changes."

Gunny nodded and made a fine adjustment to the zoom, adding in his own inimitable fashion, "OK, but this is a fucking stupid idea."

---

17 The steps needed to confirm that a Hellfire missile is ready for firing.

I suppressed a chuckle. If anything helped defuse my irritation, it was hearing from somebody who was more bent than I. Gunny's candor was another great part of this team; there wasn't a thin skin in sight. We were all passionate about our job, about making a difference in a very deadly conflict. We had a team culture where the occasional colorful language ran like water off a duck's back.

Alec called in directly, and I snatched up the handset. "Gimme something good," I said, just a bit more residual growl left in my tone than I had intended.

"Tenet and Franks are talking now. Everybody wants a direct shot at Omar."

"Well, if you can invite him to walk out in the fucking parking lot and wave, I'll be happy to oblige."

Alec took a deep breath. "I know Mark; we all know. What we got is what we got. The top guys are just getting their heads around it. The option to try and flush Omar out into the open is gaining traction."

"It has its own challenges," I said, shrugging off the last of my annoyance to go back to problem-solving mode. "Picking him out of the crowd will be at the top of the list. As soon as we confirm delivery on the first round, we'll pull out on the camera to give us eyes on most of the compound. If Omar rabbits, he'll most likely do so with a security detail wrapped around him. That should be pretty apparent."

"Yeah," Alec replied. "I'll pull in the NIMA guys and make sure they are spun up on what to look for. Roger is on station over at NSA; if anyone hits phone or radio, we'll hear it. With luck we might get a shot at Omar in a vehicle or . . ." he paused, the distaste evident in his tone, "a less culturally significant structure."

"Copy that," I replied, now pushing my best tone forward. "We're ready. Call me as soon as you hear something." Once again, the handset clicked off.

I wanted to pace, but the GCS was too fucking small. Every second we dicked around burned a little more fuel, another moment of hang time. You'd think that having no pilot in the plane meant we could run it dry waiting for a shot, but the ultrasecret nature of this aircraft, and not having a stock of spares, took crashing off the table. Calling bingo fuel, pilot jargon for "my red low-fuel light is blinking," had to happen with enough in the tank to fly Predator all the way back to our launch and recovery base. There was no pushing that envelope.

Another call came in from Alec. I reached for the handset then tapped the speakerphone instead. There was brief moment of silence, not much more than a heartbeat before I heard Alec's voice.

"We're go for the car."[18]

My instant reply was, "Put it in chat." Our chat line was another recent innovation, a text-messaging system that hit everybody's screen.

Like just about anyone in the military, I take pride in my poker face. But I'd be lying if I said that my pulse didn't jump. Otherwise, Alec couldn't have suckered me in just by hanging on the line.

---

18 Ambassador Henry A. Crumpton, in his *The Art of Intelligence*, which was cleared for publication by the CIA in 2012, discussed legal authority. Page 159 of that book reads "The White House ruled: CTC [Counterterrorism Center, part of the CIA] would keep it. But this did not end the acrimonious interagency spat about who had the authority to pull the trigger. CIA had certain lethal authorities assigned by the president, but DOD pushed back, saying that they had greater authority. The lawyers debated. Finally, policy leaders decided that CIA staff officers, at a minimum, needed to have the capability."

"Anything else?" I asked, largely to fill the void. I winced as soon as I'd spoken the words. *Don't say it, Alec. Don't.*

"Yeah," Alec replied, the smirk heavy in his voice. "Don't fuck it up." Then the line went dead.

I shook my head at having been drawn in to serve up the straight line. From the onset of operations, Alec began every mission with that directive, taking some perverse delight in the delivery. When he was unavailable at launch he had his assistant, Sharon Demcsak, pass me the word. It became our little mantra, maybe it was CIA-speak for "break a leg." Whatever the case, it struck me that, at this moment, it was the perfect thing to say.

A string of characters appeared on the chat line. "Cleared to engage the vehicle in the compound."

My eyes swept the GCS. The cramped control room was electrified. Just a moment ago, minutes felt like hours. From this moment on, I knew time would likely be in very short supply. I took a moment to close my eyes and collect my thoughts.

Had we overlooked anything? Nope. Is anything wrong? Nope. Is the hair on the back of my neck standing up? Nope. Then we're ready.

The forced humor did little to ease my tension. While success would give everyone on this team a hand in making history, failure would happen under the collective nose of the highest-ranking officials in just about everybody's food chain. "Don't fuck it up" seemed less a joke and more like profound guidance.

The words of Gene Kranz, the tough-as-nails NASA flight director who led the rescue effort that saved the crew of Apollo 13, came to mind: *Failure is not an option.* Gene was a NASA icon, the white-vested model of an old-school space-program warrior who defined his unflinching standard by two words: tough and competent. That was a standard that took man to the moon, that brought men home safe from a crippled spacecraft. It was a standard that sure as hell could drop a missile into the bed of a pickup truck. I didn't have a white vest, but . . . my eyes tracked to the Stetson on the wall. I did have the white hat.

Time to be cowboys. "Here we go," I said, loud enough to rise above the persistent din. Big turned; his eyes met mine then flicked up to the hat on my head.

The GCS fell silent. It felt like a moment to say something profound or historic, but that wasn't our style. I went with "Pilot, Sensor, you are cleared to engage the vehicle in the compound."

Nobody blinked, not a face turned away. I knew I would have seen the same thing in the mobile home next door. The eyes around me looked back with determination and focus, and it hit me just how damn proud I was of this team. Failure really wasn't an option.

I barked over the headset, "Ginger, MiG,"[19] knowing the two were in the double-wide on headsets tied to the intercom. "Lemme know what other assets are out there and what they are carrying."

I rolled through a series of rapid-fire directives, each one confirmed by terse, one-word replies before bodies scattered to their respective task. Then I turned my eyes to the MTS feed, absorbing the chessboard below.

Given the wind and target orientation, I would have preferred to approach from the southeast. That would keep the suspected mosque out of azimuth. But a wall not far from the driver's side of the car would have crowded a shot from that direction. That left us having to approach 180 degrees from where people had gone into the building—a better shot, but less ability to watch what runs out the back. Everything carried trade-offs. We settled

---

19 Captain MiG was a wicked smart analyst who specialized in air systems. A USAF Weapons School graduate, MiG was intelligent, uber professional, and had an awesome dry sense of humor.

on a northeast angle of attack as the optimum balance between visibility, hit probability, and predicted effect on target.

We had brought the plane around and descended from its loitering altitude of twenty thousand feet to a strike height of closer to ten thousand. We were shooting Kilos, or K-model Hellfires, and for all the developments in the weapon system over the years, the missile still had a brutally narrow engagement window made even more restrictive by our altitude—pretty damn high by Hellfire standards.

A pie slice makes for a decent analogy. Imagine launching a missile out from the center of the pie, trying to hit a point on the crust. A bigger slice of pie gives you a wider arc of crust, which in turn gives you a little extra flexibility in terms of how far, or from what angle, you can shoot. On the other hand, a narrow slice could leave you with just a fork-width at the receiving end.

We were looking at a damn stingy slice of pie. The position, orientation, and speed of the aircraft would all have to be spot-on.

Scott rolled the aircraft into the attack sequence, taking advantage of the opportunity to do a dry run. Gunny kept the crosshairs centered on the vehicle surrounded by men with guns and RPGs. I watched all the deltas, the variances between where the plane was and where it needed to be at the instant of launch, and counted down. *Firing in 3, 2, 1 . . . mark.* Then silence.

I crossed my arms, taking in the data on the wall of screens. The dry run felt good. Had it been an actual missile run, a Hellfire would be falling out of the sky this very moment. Big scanned the data and confirmed my gut with a curt, "That was good."

I concurred—had that been the real deal it looked to be a good shot. Nothing was left now but to do it one more time for all the marbles. We lapped back to our starting point, falling into the track we had just carved through the Afghan sky. Safeties off, weapons hot, we swept down on our target. The room became a series of last-second checks.

Laser On. Lasing. Confirm Lasing.

In a moment that seemed almost Hollywood, Scott flipped a red safety cover to reveal the firing button. Everything slowed to a crawl as he called out the final seconds.

"Launch in three . . ."

I watched the bad guys standing, guns in hand, unaware their part in this war was about to start and end in one big flash.

"Two . . ."

Eighteen months of sweat, work, frustration, all to reach this moment.

"One . . ."

"Rifle."

The tiny click might have seemed anticlimactic, but moments later, as the signal tripped a servo halfway around the world, the video feed from Predator flared black, lurching as the missile punched off the rail. The arc of flight carried the missile almost instantly out of the camera's view.

I swallowed, throat dry, and glanced at the clock. The missile would be hitting just a little over Mach 1 by now, outracing the whoosh of its solid-fuel rocket engine, then quiet, the weapon relying on gravity to further accelerate it to the point of impact.

My eyes swept the entirety of the video feed, devouring details on the HUD[20] overlays. An alarm flashed as the four box corners around the pipper, the small targeting icon in the center of the screen, failed to reappear. The sensor had broken lock on the

---

20 Heads-up display. The term has several applications, but in this context, HUD data refers to the data values overlaid onto military video footage. If you know how to read it, HUD data will tell you all sorts of things, ranging from geographic location to specific references to velocity, heading, and angles.

target. Gunny worked for about fifteen seconds to reacquire the lock before giving up, opting to guide the missile onto the target by hand. With the faintest of pressure on the stick he settled the crosshairs squarely on the vehicle.

It vanished in a black-hot flash that filled the screen. I could hear the jubilation in the double-wide, but the GCS was quiet, focused on the job at hand.

Gunny, in his deep Maine accent, broke the silence. "We have flippahs."[21]

Somebody slapped me on the shoulder, but I couldn't take my eyes off the screen as the storm of pixels resolved once more into definable shapes of gray. A pattern of wreckage radiated out from the twisted heap that just a moment ago had been a vehicle. The hulk was engulfed in gray-black flames.

It was the first shot of a revolution, a shot that would truly be heard around the world. The face of aerial warfare had just changed forever.

---

21 *Flipper* is a morbid term used in our battle-damage assessment of weapons effects on a target. It usually referred to a human target that was hurled from the area of impact by the force of the explosion. The sensor operator and analysts would always try to account for what persons were at the target before a strike, as compared to after the strike, as a measure of effectiveness. In contrast to a flipper, a *squirter* was somebody who ran away from the area of impact under his or her own power. Because of our unique ability to dwell on a target after a strike, it was important to identify targets and then track their movements for possible engagement.

2: HOW IT ALL BEGAN

ALEC BIERBAUER

AND YOU WANT IT WHEN?
January 2000

"**H**ow 'bout a big lawn dart?"

As a twenty-nine-year-old directorate of operations officer in the Central Intelligence Agency's Counterterrorism Center, I was open to using anything to identify an enemy of the United States. I just never figured I'd hear someone offer up the term *lawn dart* as part of a dead-serious discussion. But here we were in the old headquarters building at Langley, Virginia, throwing the proverbial kitchen sink on the table.

"Our clandestine sources on the ground are still our strongest tool set in Afghanistan," I countered. Call me old-fashioned, but up to that point, the most flexible, mobile, self-guided weapon in the spy business was still a case officer manipulating an indigenous network to carry out our dirty work. In support of that option we had some of the world's bravest Americans standing on the sideline, chanting "put me in coach."

As part of the government, and more specifically as part of the Clandestine Service within the CIA, we were highly averse to risking their lives for ambiguous gains. Even with a strong paramilitary division and a very capable team to challenge al-Qaeda in Afghanistan, we were hampered by sketchy intelligence and an unwillingness to put officers in harm's way. We were more comfortable hanging onto the Cold War tactics of trolling cocktail parties on the diplomatic circuit in Paris than working the back allies of Peshawar.

Mind you, the lawn-dart idea wasn't all that crazy; unattended sensors had long been a part of our surveillance arsenal. From supermax prisons to nuclear research labs, authorities monitor our most secure facilities with networks of cameras and motion detectors. Given the right sensor package, one can know in real time if something is moving in the parking lot in the dead of night and whether that movement is from a man or a stray dog.

In the intelligence world, setting up sensors to serve as a remote observation post was not uncommon. What was uncommon was doing it in the inhospitable terrain and unstable politics of Afghanistan where no natural cover, no access, and no decent source of electrical power could drive a long-term collection capability.

To complicate matters, we weren't being asked to watch a parking lot, a city, or even a county. We were hunting for one man reputed to be running around the far side of the world in the rat maze of cities, villages, goat-herder camps, and caves known as Afghanistan. That one man was Usama bin Laden.

I looked at CIA legend Charlie Allen, who sat at the head of the table. Charlie was a leader in counterterrorism before the term was even coined. Serving with the CIA since the late 1950s, Charlie had been a part of Operation Gold, the underground tapping of Soviet communications beneath the streets of Berlin. He uncovered the names of Soviet missile technicians in Cuba during the 1962 Cuban Missile Crisis and lead the National Security Council's Hostage Location Task Force in the 1980s. Since 1998 he'd served as assistant director of Central Intelligence for Collection. Called a maverick by many, Charlie was the one for a team that absolutely had to find someone or uncover secrets.

In contrast to its occupants, the conference room was nondescript, devoid of the usual collection of portraits, seals, and flags typically plastered around most government offices. A standard-issue conference table ran down the center of the room, with swivel-seating for somewhere north of twenty attendees. Another fifteen or so straight-back black chairs lined the side walls. There were no projectors, no PowerPoint decks. Instead, I sat among maybe a dozen dedicated officers flanked by some of the most powerful and innovative people I'd ever seen in one room, pulled together by nothing less than the direction of the president of the United States. I looked around the table where Charlie Allen was flanked by legends in their respective communities, guys like H. "Doc" Cabayan, Alex Lovett, and Chuck Perkins.

Doc Cabayan was the go-to problem solver for the Joint Staff at the Pentagon on technology or methodology challenges. A government civilian, Doc was the virtual gateway to all the military services and labs for hard problem-solving in support of DOD operations.

Chuck Perkins and Alex Lovett represented the Office of the Secretary of Defense (OSD) and the Advanced Systems and Concepts side of OSD. At the time, Perkins and Lovett were the civilian staff responsible for ACTDs[22] and helped make unmanned systems available in the Balkans in the mid-90s.

Lovett was like-minded in that he was determined to find a way to make things work. He was mission first and politics and organization second. Significantly, behind Lovett was a collection of

---

22 Advanced Concept Technology Demonstrations. ACTDs allow technology evaluation earlier and cheaper than would be possible through the formal acquisition of new production capabilities.

flag-waving engineers and scientists who followed him with exceptional loyalty. This collection was tucked neatly away at Naval Air Station Patuxent River under NAVAIR[23] at the direction of Special Surveillance Program and Mr. Chyau Shen. The Chyau and Alex pairing was OSD's means for getting one-off solutions to hard problems. They had reach to the best and the brightest, with an uncanny ability to break through funding and administrative barriers.

In the room from the Agency side were Charlie; Diane,[24] who was my boss at the time; and me, along with other representatives from the DS&T[25] and SAD.[26]

Taking in this assembly of rock stars, the thought crossed my mind, *What the hell am I doing here?*

The answer to that question was one part "I volunteered" and another part "I failed to duck." That's the sort of split-second decision-making that has defined or ended many a career.

The line of dominos that put me in the room was a short one that began when President Bill Clinton received a briefing from

---

23 Naval Air Systems Command. It provides material support for aircraft and airborne weapon systems for the United States Navy.

24 Diane Killip was the lead case officer in the Bin Laden Issue Station for ground assets operating in Afghanistan to target Usama bin Laden. She also led the development and implementation of covert action programs to aid these efforts. When I was detailed to the Station, I was assigned under Diane, who mentored me through a significant learning curve to transition from my DOD mindset and to understand the nuances of the National Clandestine Service.

25 Directorate of Science and Technology, the branch of the Central Intelligence Agency charged with developing and applying technology to advance the United States intelligence gathering.

26 Special Activities Division, the division of the Central Intelligence Agency responsible for covert operations, referred to as "special activities."

CIA Director George Tenet that subsequently triggered a memo to both the Pentagon and CIA from Sandy Berger, the president's national security advisor. Berger's January 2000 memo declared Usama bin Laden to be a clear and present threat to American safety, the worst threat out there. It went on to say that the United States was woefully unprepared to even find him, much less hamper his efforts.

The president had been seriously disappointed by the US efforts against bin Laden in Afghanistan and directed the CIA and Joint Staff to close that gap within nine months. The significance of the nine-month timeline was intriguing. By government standards it was an unrealistic period of time to design a new program, build and deploy it, then achieve mission success. I might be a cynic by pointing out that, if successful, the otherwise arbitrary nine-month deadline would have provided actionable intelligence on bin Laden in October 2000, just one month before the presidential general election. In my business, we are trained not to believe in coincidence.

The yoke of this effort on the CIA side of things fell squarely across the broad shoulders of Charlie Allen and Cofer Black, the latter the director of the CTC.[27] Cofer called in Bin Laden Issue Station Chief Rich Blee, a hard-charger with a distinguished lineage in intelligence services. His father, David Blee, had been honored as a "trailblazer"—one of the fifty finest officers in the history of the CIA. The senior Blee's exploits, both for the Agency and its predecessor, the Office of Strategic Services, were the stuff of covert-ops legend.

Rich was charged with leading the issue-oriented Station based in Washington, DC, to target bin Laden and his al-Qaeda network,

---

27 Counterterrorism Center, a division of the CIA's National Clandestine Service.

which had been operating largely unimpeded from Afghanistan since 1996. Rich was newly assigned as the station chief, replacing Mike Scheuer. Mike, a career CIA analyst, was arguably the smartest person on bin Laden and his al-Qaeda organization, but he lacked the operational background the Agency leadership was looking for to put his concepts into realistic plans. Rich's background of working ground operations against terrorist targets in garden zones like North Africa was well-suited for the transition and brought a more operational orientation to the Station.

The last batter in the lineup was me—*no pressure.* I had been assigned to the Bin Laden Issue Station first as a detailee from DIA and the Pentagon, then as a staff operations officer for the CIA. I didn't feel uniquely qualified to be the action officer on this requirement, but I was one of the few officers in the Station with a military background. Cofer had sent me to the meeting with guidance to the effect of "Alec, you speak Pentagon. See what you can do to answer this memo."

I snapped back to the moment as Diane spoke, shaking her head. "We've got some good assets on the ground, but they're spread too thin and are usually a day behind on anything being actionable."

Diane was a true believer in our Afghan networks, but she knew it would take something new to get the Afghans to transition from passively reporting to actually doing something about bin Laden. The authority to kill or capture him was in place, but the operational capability was still lacking. This was due in great part to bin Laden's unpredictable movements. He was rightfully paranoid and trusted almost nobody.

Valid point, I thought, nodding slowly at Diane's assertion. One of the most significant challenges with Afghanistan was the fact that it was being governed under strict sharia law by the Taliban. Foreigners were highly suspect, and the CIA was not well-staffed with ethnic Afghan-looking operations officers. As a consequence, all intelligence coming from Afghanistan was being generated by Afghan sources who were uncontrolled by Agency standards and could not be entirely trusted to provide accurate intelligence at the level that would result in orders for the military to engage targets.

The White House memo specifically required "actionable intelligence." That is an academic term for knowledge upon which one can rely to start shooting—a stiff and very unforgiving criterion. It is information from a fully vetted source that you trust implicitly. It is info that is deemed timely, not days or possibly even hours old. Things can change dramatically in hours. Once someone presses the button to launch a cruise missile, not much can be done to call it back. That person better be right when pulling the trigger.

According to US government specifications, intelligence had to be gathered by "US eyes on target" to be actionable. That meant we couldn't rely on our Afghan networks. Ideally, we needed one of our own on the ground embedded with the Afghans to put eyes on bin Laden and call for a strike. The nonpermissive environment in Afghanistan, combined with a risk-averse head shed (command center), was a virtual guarantee that "US eyes on" was never going to happen.

Aerial surveillance was the historic alternative to eyes on the ground. From the early days of biplanes, blimps, and balloons, observers with binoculars have been floating around over many battlefields involving Americans, watching enemy movement. But sending surveillance aircraft into Afghan airspace was another nonstarter.

Afghanistan is a land-locked country surrounded by less-than-friendly nations. If the United States was going to strike Afghan soil, the shot would more than likely come from the Indian Ocean. That would require overflight of Pakistan in the best of circumstances, and some ninety minutes of flight time for a cruise missile.

This was the chosen path in 1998 when the United States retaliated for the embassy bombings in East Africa. At that time, I was DIA's intelligence officer for tracking terrorist training camps worldwide. We had amassed little in the archives for Afghan training camps. Most of our focus to that point had been on Hezbollah, Popular Front for the Liberation of Palestine-General Command (PFLP-GC), Hamas, Fuerzas Armadas Revolucionarias de Colombia (FARC), Lashkar-e-Taiba (LeT), and some others.[28] When the request came in during August 1998 for bin Laden– related targets in Afghanistan, we had a handful of options, but they were for the most part little more than mud huts in the eastern mountains of Afghanistan. These offered sparse hope of hitting bin Laden.

Regardless, there had been a strong push for retaliation, so together with a number of JCS-J3[29] mid-grade action officers, we plotted point targets on mud huts for million-dollar TLAM[30] strikes.

As I looked around the room, I thought, *Maybe Cofer was right.* Aside from having previously served on active duty as an Army counterintelligence special agent, having burned the midnight oil with these notables in the basement of the Pentagon just might have left me close to qualified to bridge the Pentagon and the CIA on this particular issue. I learned a lot from these officers. At the

---

28 A group of some of the most vicious and persistent terrorist organizations, stretching from Central America to the Middle East and Asia.

29 The Operations Directorate of the Joint Chiefs of Staff. The J3 assists the chairman of the Joint Chiefs of Staff (CJCS) in carrying out responsibilities as the principal military adviser to the president as well as the Secretary of Defense.

30 Tomahawk Land Attack Missile, a cruise missile that can be fired from ship or land, with a range of up to several thousand miles.

time, we had been frustrated that we were left with mere mud huts to target. What was worse, with a philosophy of "one is none and two is one," we actually put two and sometimes three missiles on each mud hut to ensure success in the event of an errant or failed missile. That struck me as a lot of taxpayer money to blast a smoking crater into the side of some bleak mountain on the far side of the world.

Those preparing for the 1998 TLAM strikes considered that the missile overflight of Pakistan might be mistaken by the Pakistanis as a preemptive strike from India. That sort of mistake could lead to the very definition of *unintended consequences*. To prevent an undesirable shooting event between two nuclear powers, a USCENTCOM commander general was prestaged in Islamabad to deliver a well-timed notification to Pakistan's leaders when the missiles were launched.

The intelligence from the CIA's Afghan network suggested that bin Laden was at the Zhawar Kili al Badr camps outside of Khowst when we started the strikes. Because of poor coordination between the CIA and Pentagon, one of the network's subsources was at the camp when the missiles arrived. He was killed along with a number of legitimate al-Qaeda targets.

Most reports indicated that bin Laden had left the camp prior to impact, but it remained unclear if he had somehow been warned. Like many others at the CIA and DIA, I was convinced of the unholy trinity between Pakistani intelligence, the Taliban, and bin Laden. There was ample reason to believe that any notification given to the Pakistanis would be relayed immediately to bin Laden, likely through the Taliban. In a world of global communication, news can travel faster than missiles.

But the prospect of Pakistan believing it was being attacked by India was too high a risk to not give them advance word. The prospect of a war between India and Pakistan is chilling enough without adding the moral weight of inadvertently being the catalyst. The insult upon injury was that entities like al-Qaeda would doubtlessly grow in those ashes and prosper in the subsequent instability and chaos. As second-order effects go, kicking off a nuclear war in Asia was certain to have some severe career repercussions.

That left us having to work within some immoveable left and right boundaries. The mission at hand today was to figure out how to surgically excise the cancer that was metastasizing in Afghanistan's training camps.

The proposed mission was fraught with trade-offs right out of the gate. In order to meet the objective of the memo, we'd have to make some hard planning assumptions and accept some uncomfortable constraints. The latter practice was no more popular in the halls of the Agency than it was in the Pentagon. I looked down at the printed list on the table in front of me and counted six that we had drafted before the meeting.

Assumption One

Sandy Berger and President Clinton thought the CIA technical and covert action programs in Afghanistan were sorely lacking, which prompted the memo. When George Tenet briefed President Clinton, we had a number of technical and clandestine-source initiatives ongoing in Afghanistan, both network and surrogate options, but neither were expected to yield bin Laden or actionable intelligence as to his whereabouts.

This wasn't entirely a knock on our efforts. With the Taliban in firm control of probably 80 percent of Afghanistan, the majority of Afghan civilians were living under heavy pressure. The Taliban had no concern for pesky things like human rights or UN resolutions. If they chose to make a point, they could shoot men, women, or children in the head. That makes for some strong control of a civilian population.

To complicate matters, illiteracy was rampant across almost all of Afghanistan. Simple technologies such as cell phones, cameras, or GPS devices were virtually unheard of; most were effectively banned for everyday Afghans. For CIA technical services, this created a tremendous problem. Electronic collection tools are often hidden inside host devices, such as a cell phone, an item so prevalent as to be invisible in Western culture. But the average Afghan not only lacked a camera, but also had likely never seen one nor understood how to use it. Imagine having to educate an asset on what a camera is before you teach him to use it, conceal it, and actually gather critical and timely information in a deadly environment. That's what we in the trade call a nonstarter.

The previous year of technical advances for putting such tools into Afghanistan had yielded one concealed camera in the hands of a fairly capable and willing asset, but despite his commitment and intentions he was never able to get the concealed camera in proximity to bin Laden. The pictures we did receive were useful but were of lower-level al-Qaeda personalities. By the time the photos were exfiltrated out of Afghanistan, most had become stale in terms of value.

Assumption Two

Growing a better or more capable collection network or surrogate force was not a viable option within our time constraints. This was a fair point; we couldn't develop new networks overnight. Gathering credible, actionable intel takes time, and doing so in a rush would be a long shot of epic proportions. What sources we had in Afghanistan were poor and almost impossible to corroborate. Data points were typically unsupportable and after the fact. Given DOD's targeting requirements and the gaps in the CIA's networks, we knew source reporting alone wasn't going to get us there.

Assumption Three

The US government couldn't muster the temerity to put manned aircraft over Afghanistan or troops on the ground after the embassy bombings in 1998. That being the case, it was unlikely the government would have the resolve to do so now. Putting US personnel in Afghanistan carried huge consequences and was viewed as a wholly unacceptable risk. Afghanistan is land-locked and with no sufficiently friendly neighbors, and we couldn't fulfill the search and rescue or support requirements needed to send personnel over or into Afghanistan.

Since August 1998 there were ebbs and flows of interest and commitment to getting serious about getting bin Laden. The dependency on the reporting from the CIA source network became a daily thirst at the Pentagon as the J3 planners worked through options for what would range from small deniable actions on the ground up through an all-out US invasion of the country.

A victim of our technologies and bureaucracy, the US government set some pretty high standards for the hunt for bin Laden. The dependency on the reporting from the CIA source network was to find the elusive credible and timely intelligence. Thus far it had not materialized, or we failed to realize what we had and act on it decisively.

Among those to champion the cause was US Army Major General John Maher. Maher was a solid push behind initiatives for boots-on-the-ground options in Afghanistan in the wake of the embassy attacks in 1998. Sadly, Maher ran up against a lack of political will and interest in pursuing this option. The challenge was not simple. Numerous complications hindered serious consideration of a boots-on-the-ground option, such as specific and timely intelligence, third-country access, CSAR capability,[31] a capable force, and a receiving entity. Maher was one of the few in the Pentagon approved to view the CIA raw source reporting; over time, he became very good at qualifying the reports and knowing the ins and outs of the personalities discussed. As the deputy J3, he had taken a keen interest, if not an outright obsession, in the bin Laden targeting problem.

One of his subordinates—the J33 current ops director—was US Army Brigadier General Bob Wagner. Wagner was a no-nonsense officer who was far more skeptical than most on the CIA reporting; he was not at all shy about challenging the reports and those of us charged with giving them context.

Specific and timely intelligence demands that if all other factors are present and acceptable, we know where the target is and have near real-time knowledge to make the operations worthwhile. In the intel/ops world we also want this intelligence to be as close to perfect and reliable as possible.

Assumption Four

A technical solution was our charter from the memo, and given the timeline we had to set some conditions. This meant a focus on proven tech—no drawing board ideas, no beta test products, and nothing that hinged on putting an American in harm's way. In the engineering world, the progress of technological development is

---

31 Combat search and rescue, a vital contingency ability to quickly enter enemy territory to rescue a downed pilot or people stranded behind enemy lines or in imminent danger. CSAR usually involves helicopters and highly skilled operators, prepositioned as close as possible to the operating environment. This was especially challenging in Afghanistan, due to the land-locked and neighboring-countries reasons mentioned earlier. The Indian Ocean was too far away to offer a timely ship-launched response capability.

measured on a scale of 1 to 9, dubbed "technology readiness level," where 1 is a cocktail-napkin sketch and 9 is a working product. We could accept no less than TRL-6, a full-up prototype. A 7 or 8 was preferable. According to top-level restrictions we needed an immediate, reliable technical solution, free from risk to US persons, without significant cost, and with limited to no political risk. Typically those are filed under *U* for unicorn.

Assumption Five

If by some sequence of miracles, we actually came up with actionable intelligence, we fully expected action to follow. If we point out a fly to the guy with the flyswatter, the next thing we want to see is a splat mark on the wall. That sounds self-evident, but in a world of shifting political agendas, this one carried some of the most undefinable risk.

It also imposed a severe operational time constraint. The critical INT would need to be imagery; it had to be delivered in real time and would need to be triggered by a timely HUMINT[32] or SIGINT[33] event.

---

32 HUMan INTelligence. HUMINT encompasses knowledge acquired through interpersonal contact, typically through a network of trained collectors and assets developed across the world. Due to the highly subjective nature of HUMINT data, every assertion must be vetted and corroborated before it can be considered reliable.

33 SIGnal INTelligence. SIGINT encompasses knowledge acquired through the interception of transmissions over radio, internet, satellite connections, and so forth. When further specificity is needed, SIGINT can be subdivided into separate components such as COMINT (comms intercepts) or ELINT (electronic intercept). Because a substantial portion of transmitted information is encrypted for security purposes, SIGINT functions closely with the science of cryptography, focused on breaking a wide range of codes and encryption strategies.

Another vital arrow in the intelligence quiver, SIGINT encom-passed an arsenal of high-tech eavesdropping technologies that day and night plucked information from phone lines and radio waves, even snatching data bits as they bounced from earth to satellites in space. A staple of the Cold War with a focus on industrial super-powers, SIGINT had been both empowered and challenged by the explosion of personal-communication technology.

Yes, some of the bad guys used electronic comms, but so did everybody else on the planet. Even known threats had become nee-dles in a global haystack. Although we could focus on almost any point in the signal-sphere, we couldn't watch all of it all the time. For all its magic, SIGINT would not pluck the positive identifica-tion we needed out of the haystack. We had come full circle, back to the need to put American eyes on the target.

Assumption Six

If we were to have a prayer of actionable intelligence prior to November 2000, we would need to be fully operational no later than the end of summer, at the very latest. If the clock hand had revolved any faster, it would have been a fan.

Absent the ability to put boots on the ground, some form of technology would clearly have to fill the gap in finding a man mov-ing randomly through Afghanistan. That's where the lawn darts first came in. Unlike securing a facility at home, you don't get to send technicians into a hostile nation and mount carefully aimed cameras on poles. Technical surveillance had to rely on a class of gizmos called UGS, or unattended ground sensors—devices that could be chucked out the window of a passing car or dropped by aircraft in the dead of night. These ran from very simple widgets to the stuff of Jason Bourne movies.

A simple example would be little more than an average micro-phone, an omnidirectional seismic sensor that listens for the vibration of a vehicle passing within a certain range. We dress it up to look like a rock and chuck it along a roadside with little care which way it ends up pointing. But a seismic sensor has limits. It can tell us when a vehicle drives by, maybe even give us an idea of size and speed of travel, but it cannot confirm make and model, cannot tell us the vehicle's color or the license-plate number. Most important, it can't tell us who is riding in the front seat.

Mind you, even simple sensors consume power, both when collecting data and, more important, when transmitting it back home. How far it can sense and for what duration, and how it phones home are all parts of a very intricate balancing act involving terms like volts, ohms, and amps. Nothing is simple about making the simplest sensor work in the field. SWaP, or size, weight, and power, are the critical trade-offs that need to be made depending on how long we need a device to operate, how often we need it to transmit, and how undetectable it must remain.

That last point is critical. Like any piece of the spy game, sensors have to live and work not in a lab or a test range but in the backyard of an enemy who doesn't want to be observed. On top of all the limits imposed by physics are the threats of enemy detection and countermeasures, of freezing cold or blistering sun, or of being chewed on by a goat.

The challenge on a UGS strategy is multiplied a hundred-fold when we add the need to identify one human being in real time, potentially among a crowd of other people. Vibration or heat can only tell us when to look, but if we want to recognize a man we need to see him.

For that we need a camera, which needs to be pointed. But a sensor dropped from the air can bounce around like a pinball on landing, riding Murphy's Law into any nearby crevice or pothole. On ninety-nine of a hundred tries it will end up stuck behind the rock, not perched on top of it.

To put that last challenge into perspective, take out a scrap of paper and try to sketch a GoPro housing you can drop from a plane. It has to survive the impact and then be able to figure out where it is relative to the target. It needs to crawl, jump, or fly around any obstruction, then park itself pointing at a meaningful door, gate, or curve in the road. It has to remain camouflaged through the whole process, and every erg of power used to position the device eats away at its useful lifespan. And while you are at it, make it run for a month from a couple AA batteries.

The lawn-dart idea was a response to that problem, a sensor that could "stick the landing" like an Olympic gymnast. But from the rugged Hindu Kush mountains up north to the rocky deserts of Helmand, soft, fertile soil was in short supply. A lawn dart of any shape or size would be no more likely to stick in much of Afghanistan than it would in the concrete expanse of Manhattan.

Three hours of brainstorming crawled by, with little more to show for that than a list of disjointed components as the best of bad options. We had been beating to death the desire to create an echelon approach, augmenting our Afghan sources on the ground with sensors that could characterize and alert us when a vehicle—or better yet, a convoy of three or more vehicles—approached one of the dozen or so locations bin Laden visited. If a sensor pinged, we could target our sources on that location and have cameras in sensors provide additional clarification. But all this layering still left us at a loss for putting US eyes on target.

Like everybody around the table, I shoved increasingly disjointed puzzle pieces around in my mind. We couldn't put an American on the ground to observe firsthand. We couldn't send an American pilot into enemy airspace. We didn't have sci-fi robots that could crawl across the ground like a terminator and watch from a nearby ditch. The conundrum made my brain hurt.

We had a promising concept from Alex Lovett to use the terrain of Afghanistan for as much standoff as possible and try to smuggle in some high-powered optics. This plan would require some research and analysis, but its application would obviously be limited in where we could put the optics. And there seemed to be some SWaP issues associated with running them.

How the hell do we park a camera—an unseen camera that we can reposition at will—within eyesight of our target so it can beam a signal back to US eyes sitting at home?

Some moments in life you hear words being spoken, and you pray to God they didn't just come from your own mouth.

"If we can't send a pilot into harm's way, can we just send the plane?"

The room fell silent, and it wasn't that warm, satisfied silence when we all sit back and admire a job well done. The suggestion sat like a steaming turd in the middle of the table, and nobody was stepping up to claim it.

Charlie looked down the length of the table and fixed on Lovett. "Is that possible?"

With remarkable poise for someone who just had the hot potato dropped in his lap, Lovett started slowly: "Yeah, an unmanned aircraft. A flying camera." The concept felt like it had merit, but it was in dire need of a hot shower and some quiet reflection. Neither was about to happen, so Lovett ran with the ball. If the DOD wasn't willing to put manned surveillance over Afghanistan, some sort of model airplane was the next best thing.

The idea began to snowball. "We pilot it remotely, beam the picture back to the ground where US eyes can watch in real time. In a sense, it is eyes on target—that ought to be enough to satisfy the decision-makers."

Despite the wind-up-toy first impression, the appeal started to grow. I saw gears turning. But just as prominent was a teeter-totter between skepticism and optimism. I scanned the various faces, categorizing the different looks anywhere from "Tell me more" to "It's been nice working with you."

Charlie's eyes held neither; his expression was deadly serious. He turned to me and asked, "You think we can do that?"

My brain chewed through details. There was a good chance the idea would fit inside the planning assumptions, but to my knowledge no one had ever given the idea serious consideration. We'd have to work through the technological and political consequences: how are we going to remotely fly an aircraft, penetrate sovereign territory of another nation, not run into commercial aircraft, and find one man in a country the size of Texas? Like any good soldier with an underdeveloped sense of career survival, I looked back at Charlie and said, "Yes sir, I do."

The next thirty minutes were a blur of challenges and suggestions, of questions and answers, flying in from all sides. From the moment that the idea of an unmanned aircraft finally took hold, everybody around the table began to offer insights, highlight risks, suggest mitigations. With the president's clock ticking, we mapped out a short-run timeline to separate science from science fiction. The first step was to find an unmanned airplane.

MAD SCIENCE
January 2000

In Mary Shelley's gothic novel *Frankenstein*, crazed Victor Frankenstein labored in the dead of night to stitch together a living man from the mismatched parts of dead ones. As I looked around my office at a mural of catalog pages, sketches, and blueprints that covered a desk, a table, and several walls, I was struck by a thought: *Victor was an amateur.*

Despite his macabre mission, the good doctor had the freedom to harvest all his basic parts from the same species. He didn't depend on other people or on organizations with their competing agendas and budgets. In stark contrast, the sheets of paper now displayed around my office harkened from every corner of the battlefield: spec sheets for several different airplanes, a brochure on ground antennas, an article on a converted automotive trailer, and a SATCOM[34] package that looked like the guts of a broadcast news truck stuffed into a man-portable pack.

Taped to the walls were pages photocopied out of a volume of *Jane's Flight Avionics* or torn from an issue of *Popular Science*. The mix included radars and radios, disparate parts that had to not only work on their own but also play nice with one another in ways nobody ever envisioned. Manufacturers promised the moon in terms of performance, but every pledge was tempered with the caveat "your mileage may vary."

As if all this wasn't complex enough, my outside voice had once again managed to move faster than my common sense during the last brain-storming process, so we added the desire to push the video not just to where the pilot was on the ground but also all the way back to the decision-makers—a live stream that could be viewed inside the DC Beltway. I grumbled at my own enthusiasm, thankful I hadn't contemplated teleporters or cloaking devices along the way.

The chasm between what was promised and what was likely grew. The whole process felt like shopping for body armor when the distinction between "bulletproof" and "bullet-resistant" suddenly

---

34 SATellite COMmunications. Unlike cellular communications that typically rely on line of sight to a tower, SATCOM beams directly to the constellation of satellites orbiting overhead. As such, SATCOM can be used to largely "place a call from anywhere." This is notably vital in rugged or remote terrain.

ceased to be a matter of semantics. If this was going to work at all, we desperately needed the no-shit ground truth on a great many parts before our spliced-together monster could take life. That meant hands-on tests, not brochures and demos.

Lovett, my primary point of contact at the office of the Secretary of Defense, was uniquely equipped to sort through an arsenal of technologies. Through his high-level OSD position, Lovett had access to damn near all the DOD resources as well as the national labs and contractors.

As a bonus, that position allowed us to ask questions without tipping the Agency's hand. Admittedly, that hand wasn't much of a leap of logic the moment we started discussing hypothetical missions to penetrate nonpermissive airspace to hunt hard-to-find individuals. When that happens you just have to settle for plausible deniability—that or the old "if I told you I'd have to kill you."

Central to our emerging Frankenstein was the Air Force's little-known airframe called Predator, built by General Atomics. After some initial issues with the Army during a two-year Advanced Concept Technology Demonstration, the Predator was turned over to the Air Force in 1995. It had performed well. Its early success in that series of tests led to Predator being deployed to the Balkans in 1995. The aircraft flew overt observation missions out of Gjadër, Albania, under a program name Nomad Vigil and later out of Hungary for various Balkan operations.

Predator went on to see action as a reconnaissance platform in Kosovo in the 1999 Operation Allied Force. Yeah, the Predator planes suffered a few crashes and shoot-downs, but overall they performed admirably. Still, flying a remote-controlled airplane in line-of-sight conditions was a far cry from pushing live video from the far side of the planet. If we were going to send it off to hunt bin Laden, we had to find out if Predator could step up its game.

That took Lovett and me to Rancho Bernardo in San Diego, California, to meet with General Atomics. Manufacturer of both the Predator and its little brother, IGNAT, General Atomics was run by retired Vice Admiral Tom Cassidy, a former Navy test pilot.

Lovett had a history of working with Cassidy back when Predator was just a science project. That brotherhood was strengthened by a shared Navy heritage. As such, when we met with Cassidy in his office, I put it to him rather plainly. I said we were interested in using his system to penetrate sovereign airspace without permission and to hunt for a high-valued target.

He took it better than I expected. While the admiral was all about greater interest in his platform, he was actually hesitant to have it associated with this sort of project. He said that Predator was really not suited for such clandestine work and that he was uncertain how it would perform to the outlined mission. He suggested it might serve better in more of a declared and conventional fashion. For him, the prospect was fraught with undefined variables. A failed program would have a negative impact on future sales and the platform's reputation.

I noted his thoughts, adding them to the growing list of problematic inputs that would need creative context development before briefing them back to the boss. At this point we were not married to Predator or IGNAT; in fact, our technical solution might not involve a UAV at all. But polite bullshit aside, the Air Force owned the platforms, and if we ultimately settled on Predator we would be using it regardless of GA's consent. We would need its subject-matter expertise, however, along with assistance with some critical modifications.

As the discussion progressed, we dug into the nitty gritty of how the platform would perform at the margins of its advertised range and endurance, as well as with a greatly reduced maintenance and operator footprint. Cassidy was on board and helpful but still an admitted skeptic.

Given that we were going to try to hide from commercial and military airspace controllers, we needed to know what Predator looked like on radar. Cassidy answered with confidence that it would appear like nothing more than a "flock of geese." With our limited time and budget, we decided we would take that assertion as gospel—at least until proven otherwise. We would find out for ourselves soon enough.

Another important loop was closed during the meeting. I had heard "Big Safari" during the initial meeting with Charlie Allen, but frankly it had been lost in a sea of new acronyms. I had it checked as a follow-up item. Though Cassidy's skepticism had me a bit concerned about the company's level of commitment, he noted that the Air Force had an element called Big Safari parked literally inside of GA. Its job, he explained, was to oversee unique advances on behalf of the Air Force.

That was perfect—if it saw the merits of the program we would have an advocate inside the factory, mitigating anxieties about GA's marketing and future. I learned that Big Safari was headquartered at Wright-Patterson Air Force Base in Dayton, Ohio, under the leader-ship of Mr. Bill Grimes. I had come across Bill's name during initial research but only knew him by reputation—a legend in the world of skunk works[35] with a reputation for achieving technical miracles on a regular basis. Bill's local lead at General Atomics was Major Brian "Radz" Raduenz, a uniformed Air Force officer who, if I had my way, was about to have all his vacations and weekends put on hold while we tried to snag a couple of those technical miracles for our interests.

---

35 Originally the official pseudonym for the Lockheed-Martin Advanced Development Program, which produced advanced aircraft like the SR-71 Blackbird. The term "skunk works" evolved over time to encompass a range of secret projects where brilliant engineers performed acts of technological wizardry at facilities far from the public eye.

At the close of our visit, we toured the manufacturing facility and had the opportunity to meet some of the other people who might ultimately support the program. I would have bet my paycheck that not a one of them thought the anonymous suits from Washington were weighing the option of deploying them under the cover of the CIA in just a matter of months.

Almost as an afterthought, Cassidy paused the tour to point out a small rack of computers with a chair attached. It was another little science project the company was playing with, he explained, to build the minimum footprint for operating a Predator from inside the confined space of a submarine.

I was impressed with the bare-bones approach, a stark contrast from the normal control package that filled a twenty-foot sea-land container. From what I knew of the Air Force, an MTOE[36] for this sort of effort would likely include some 110 people and a couple C-5 Galaxy aircraft worth of equipment. Coming from the world of covert operations, I was thinking more along the lines of two burly guys and a truck.

Staring at the compact control station, the burgeoning mad scientist in me began to imagine how yet another organ donor might add to the whole that was our Frankenstein tank-heli-plane. "What the hell," I muttered to myself, "let's throw a scoop of submarine into the mix."

THE VEGAS FLY-OFF
March 2000

We passed a small casino and a dusty old trailer park before we swung into the gates of Indian Springs Air Force Auxiliary Field.

---

36 Modified Table of Organization and Equipment, a Defense Department document that prescribes the personnel, equipment, and consumables that will make up an operational package.

Although its roots trace back to WWII, Indian Springs had been relegated to "divert field" duty—a place to land if you couldn't get to where you really wanted to go. In its day it had served as a Red Flag aerial combat training airfield, had a stint as a Special Weapons Center site, and provided winter training for the Thunderbirds aerial stunt team. But while technically US military property, Indian Springs was on the edge of being considered a remote assignment.

It was late March, but the southern Nevada heat was already leaning into summer. I was driving the car that carried Cofer Black and Rich Blee. Just off my tail was another car with General Atomics's Vice President Frank Pace, along with Roger Cressey, director of transnational threats at the National Security Council.

I had carried the results of our initial research back to my bosses in the CTC. My formal recommendation to hang our hopes on an unmanned aircraft was greeted with no small measure of skepticism. I had asked a now predominately human intelligence-gathering agency to propose a largely experimental aircraft as the solution to the largest manhunt in modern history. As they say in the South, "That just ain't how we do things around here."

While unhappy with a technical approach, Cofer said before we went outside the CIA for a solution we would need to exhaust all possible internal technical capabilities, to include the IGNAT 750. The IGNAT was our own unmanned system, managed by Air Branch inside the CIA's paramilitary group. I ran down the office that owned the IGNAT and reported back to Cofer with a side-by-side briefing chart on IGNAT versus Predator. It was clear to me that we needed the DOD system. To Cofer, the politics of organizations would need to be factored in. When it came up that both systems were housed near Las Vegas, Cofer called for a fly-off.

In addition to being an auxiliary airfield, Indian Springs was a training location for the Air Force Predator program. Down the road from the trailer park was an old prison, one more notch for a low first impression.

Not too high a bar for competition, I thought with smugness. *Surely, the Agency airfield can compete with this.*

Though no shots would be fired in this little head-to-head, today was to be a dogfight of sorts, a competition between two planes and two services: the US Air Force versus the CIA. The Agency had a background of involvement in experimental aircraft development, with highlights that included the science-fiction-looking YF-12A, which evolved into the iconic SR-71 Blackbird.

Today's fight would be pretty much like any other aerial contest, setting aside the fact that the planes would fly at different times and in different places and that neither one would have a pilot inside.

We were met at the door by Air Force Captain Scott Swanson and escorted to a conference room where we were briefed on the events taking place. It quickly became apparent that little to nothing of what we were about to see was being staged for our benefit. The Air Force was actively testing and training on the Predator, and today's exercise was more "just another day at work" than some sort of mad science experiment. The operators knew their jobs, and people went about their business in a professional manner.

After a canned capabilities briefing with the obligatory PowerPoint slides, we were taken out to a large fabric-covered clamshell hangar where a working Predator was parked. A cluster of maintenance guys stood off to one side, clearly waiting for us to get out of the way so they could get back to work. I'd seen the aircraft in the factory, but here in the hangar, under a bright blue sky, the impact was altogether different.

Good God, that sucker's ugly.

A bit shy of thirty feet in length, its windowless nose was wide and domed, with a prominent ball on the chin that gave it a distinct underbite. The two-bladed prop was stuck on the wrong end, just behind tailfins that drooped like they had deflated. In an age where sleek and swept was the mark of fine breeding, her scrawny wings stuck out straight to either side.

I reminded myself, *We aren't picking a bird for fighting, just one that can fly in a circle.*

We finished running our hands over white-painted skin and walked out to the runway talking tech. With a cruising top end around a hundred miles an hour, Predator certainly wasn't going to chase its prey, but with a stall speed down around sixty-two miles per hour and boasting about twenty-four hours of flight time, it could hang over a location like a kite.

I heard a sound coming in from my right, a harsh, raucous chainsaw howl as another Predator came down out of the sky. I'd been around my share of jets, and by comparison this descent seemed to take forever. Seconds ticked by, and the plane was still hanging in the air. It was louder but didn't seem to be a whole lot closer. When it finally flew by—low and painfully slow—I could read the stickers on its fuselage. It struck me that today was going to be less of a dogfight and more of a lawnmower race.

After watching the plane crawl back into the sky, we entered the GCS, set up in what looked like an old NASCAR trailer. Wedged inside was a video-gamer's wet dream. Two chairs sat side-by-side, each equipped with joysticks and switch panels amid a series of flat-screen TVs. Charts and clipboards filled the gaps. It looked like what was left after peeling the rest of the plane away from the cockpit. We spoke with the flight team, a pilot and sensor operator, asking the kinds of questions that aren't printed on a spec sheet.

As we walked back to the car, there was a good discussion and a fairly positive feel about what we had seen. Predator was technically shaping up as a viable platform. The major reservation was that it involved the conventional Air Force, and nobody at the Agency saw that as a good marriage for a high-stakes intel op. The Air Force could certainly be clandestine—it had a rich history of daring and precision in the world of top secret and a history with the Agency doing reconnaissance. But we saw this as a cutting-edge CIA mission using our unique authorities, and to us that took things to a different level.

Our technical experience with the Air Force demo set a higher bar than I anticipated for the CIA to follow. As a matter of natural competitiveness, I was anxious to see how my home team matched up. I was okay with Predator winning—in fact, it seemed like the best choice on paper—but the Agency in me wanted to make the fly-off a good fight.

In the CIA corner was the IGNAT, actually a "kissing cousin" of the Predator. Envisioned as a target drone, the IGNAT started out as the brainchild of Abraham Karem, a chief designer for the Israeli Air Force. Karem's company went bankrupt, only to be bought up by a defense contractor that sold five aircraft to the CIA. We were on our way to see one of them.

Gaining access to the facility where the IGNAT was warehoused was no easy feat. Perimeter security for the installation was significant. Once we were on the facility, though, a long, empty drive took us to a remote back corner of the property.

I cringed when our GPS led us to a ramshackle gate that hung open as part of a chain-link fence with no sign, but I told myself that it would be poor tradecraft to hang a placard that read Top Secret CIA Aircraft Facility.

Turning in, we followed the poorly maintained road to a lone hangar surrounded by an odd collection of fifty-five-gallon drums, tires, and cast-off flight-line equipment. The place looked more like a parts graveyard than an op center. And there in front of the hangar door, parked in a folding lawn chair, was Earl.

To say that Earl wasn't ready to do a demo would be to suggest that Earl knew there was supposed to be a demo in the first place. Seemingly puzzled at having a visitor in his corner of the empty desert, he was all too happy to pull the dust-ridden tarp off the IGNAT, making polite excuses for the various parts that weren't currently bolted on. I didn't need to be Werner von Braun to figure out this bird wasn't going to fly today, tomorrow, or anytime in the near future.

We collected what data we could walking around the plane, asking whatever questions we thought Earl might be able to answer. But it was painfully clear that today had ended up less a fight than a forfeit. To me, even a head-to-head fly off wouldn't have made a difference in the technical ability to meet the requirements for this mission, but we needed to go through the motions and our due diligence. If we were going to put a virtual "American eye" on bin Laden in the next eight months, that eye would be flying on Air Force wings.

STRANGE BEDFELLOWS
April 2000

As a general bit of career advice in the intelligence community, should you find yourself in the Langley headquarters briefing the director of the CIA's Counterterrorism Center, don't lead with, "Sir, I know what you want, but you can't have it."

There was a great deal of institutional pressure in favor of selecting the IGNAT. Whereas America has a history rich in accomplishments great and small, rarely can examples be found of different organizations playing nice together in the sandbox—especially organizations on opposing sides of the military and intel-community chasm. But given all the technological hurdles still ahead of us, it was clear to me that Earl wasn't going to be a part of the solution.

"It doesn't work without SATCOM, sir." I chose to sidestep the institutional comparison and instead hung my hat on specifics of the airframe. "Its line-of-sight limit is about 150 nautical miles from the airfield, and we would need more on the order of 500. Plus, we have to bounce a signal halfway around the world. The Ku-band SATCOM alone runs almost seventy-five pounds, and the IGNAT can't get off the runway with that much weight, nor does it have the onboard power to support it."

The argument wasn't winning any friends, with opponents that included Special Activities Division. That division had come away from the aborted fly-off with something of a bloody nose. But nobody could fault my logic, nor could anyone suggest a way around the SATCOM limitations.

When all was said and done as far as decision-making internal to CTC, I was selected to carry the message to the chief of SAD. Since it was my idea to break with Agency tradition and use Pentagon assets, I could deliver the news. We were still going to need some of SAD's people and equipment, but their unmanned capability wouldn't make the cut.

I once again fell back on my technical argument to explain our decision to the chief of SAD. His response was "so you are basically asking me if you can have the keys to my car to take my wife out on a date." I guess that about summed it up and set the stage for what would continue to be institutional resistance.

I expected to be happier when the final decision was made, but the approval of Predator felt less like a victory and more like that moment when the alternatives fell over dead—much like my career would fall over dead if we came up short. I'd positioned myself against some of the Agency's heavy hitters who had gone on record as opposing my recommendation. If this tanked, no doubt my tombstone would read SHOULDA GONE WITH THE IN-HOUSE SOLUTION.

Once that decision was made, however, it was down to business. Ultimately, the Predator we hoped would fly would likely have little resemblance to the aircraft running laps in Nevada. The array of modifications would depend on numerous mission parameters: how far we had to fly, at what altitudes, and so forth.

Those questions launched us into the fuzzy realm where physics and politics chafed one another. Although we had a reasonable shot at bouncing a TV signal around the globe, the actual flight range of the aircraft was far, far less. Since we couldn't take off and land inside Afghanistan, our options would have to be found at the overlap of performance and permissibility.

Iran was not an option. We had reasonable belief that throughout the late 1990s there had been ongoing dialogue between Iran and bin Laden. From a targeting perspective, Iran was to the west while we were focused on the east side of Afghanistan. Those realities, combined with the lack of any diplomatic or intelligence exchange with Iran, scratched it off the list immediately.

We looked at the remaining nations that surrounded Afghanistan. They were a mixed bag of good and bad relationships, geographic challenges, and political minefields—not as bad as Iran, but close.

Ultimately, we only had one friend in the neighborhood.[37] Some might argue that Uzbekistan wasn't so much our friend as that it disliked some of the same people as did the United States.

---

37 CIA Director George Tenet, in his book, *At the Center of the Storm*, which was cleared for publication by the CIA in 2007, states on page 177: "We told the president that our only real ally on the Afghan border thus far had been Uzbekistan, where we had established important intelligence-collection capabilities and had trained a special team to launch operations inside Afghanistan. We knew that Uzbekistan would be our most important jumping-off point in aiding the Northern Alliance."

Nothing in my experience clearly established that one criterion was any better than the other. Whatever the case, we had an aggressive ambassador and chief of station there, which could help selling our crazy idea on both sides of the globe.

That call left us with just two major resources to be nailed down: a place to work and the people to staff it. Given the ad hoc nature of everything else thus far, it seemed no time to break with tradition. We were relegated to the Global Response Center in Langley, perhaps by no small coincidence located just down the hall from Charlie Allen's office.

The GRC had been built as a contingency, a full-blown CTC command center on standby in case some horrific act of terrorism took place. The dark, muted interior had that Doctor Strangelove sort of feel, the main room dominated by an odd-shaped conference table that was in turn surrounded by a number of big screens.

Given that no major terrorist events had taken place recently, the GRC was available. It was a great facility but one that was specialized in ways that few teams really knew how to leverage. I was tossed the proverbial keys with a shrug and a "see what you can do with it."

If that sentiment applied to our location, it doubled down on our personnel. Every organization has them, anomalies who possess qualities that seem ideal on one hand but prove unmanageable, or incomprehensible, to managers happiest with cookie-cutter employees. Those were just the sort of guys I was looking for.

Dave Phillips,[38] or "Doc" to his friends, was something of a Zen master of a global game of Where's Waldo. Seen by some to

---

38 Dave Phillips was contracted to the Bin Laden Issue Station as a part of the team in the late 1990s to help build covert action programs to defeat or degrade bin Laden's abilities. His medical background was especially relevant to the capture plans worked in 1998–99. Dave's passion and commitment under Diane's leadership produced the best compilation of known and suspected bin Laden movements in Afghanistan.

come in day after day and perform arcane rituals over otherwise inscrutable maps, Doc's skillset ran down the centerline of an effort to find one man. As a former Special Forces medic, Doc had been a part of prior planning to deal with bin Laden. In the late 1990s, there was a fairly mature plan on how to handle the possibility of his capture. That plan would have squirreled Doc and bin Laden away inside a shipping crate to secretly move the high value target as cargo, below the proverbial radar of customs services anywhere the plane needed to stop on the way home. We joked that it would have been the original "Doc in a Box."

A gift from the NSA, Roger was another "one of a kind"—the only Top Secret/Sensitive Compartmented Information-cleared Pashto linguist with access to high-end SIGINT platforms. With the ability to divine not only literal meaning but also the subtleties of nuance, Roger had the golden ear that we hoped would complement Predator's unblinking eye.

My primary agency partner in crime was Hal, a gruff former Navy SEAL who was in Alec Station[39] as a contractor providing his special-operations expertise to all planning against the bin Laden target. We balanced each other well. As we ran into one challenge after another inside the Agency, I would work the diplomatic approach with leadership, and Hal would intimidate and coerce the appropriate action officer until we achieved our desired effect. This good cop/bad cop approach served us well as we successfully begged, borrowed, and stole the necessary components for our science project.

---

39 Most Agency Stations are defined by geography and the country they address. As a break from this rule, Alec Station was an issue-oriented CIA station created to target bin Laden and al-Qaeda anywhere in the world. Alec Station was based in Washington, DC, and was dependent on coordination with the other geographic stations for support.

On my first visit to CTC's Alec Station in 1998, I was going to meet the person who stood at the forefront of managing all programs against bin Laden in Afghanistan. Expecting a weathered, snake-eating veteran of wars and secret conflicts from around the globe, I was shocked to encounter Diane Killip, a woman in her forties who initially reminded me in appearance, tone, and demeanor of Glenda, the good witch from *The Wizard of Oz*. But one would be in error to make assumptions based on her outward demeanor; Diane boiled with energy, passion, and commitment to take down the evil lurking in the world—with her own hands if given half a chance.

Diane and I had numerous surreal conversations that transitioned from an appreciation for the art in her office or her teenage daughter's scholastic accomplishments, to the urgent need to shed bin Laden's blood and the numerous ways we could inflict the first cut. "My daughter got her semester grades last night, and she is doing very well. She is starting to look at colleges, and I think we will visit the College of William and Mary this weekend. Oh, and our asset on the ground needs to build the car bomb so it looks like the Russians or Iranians built it. Let's tell him to build a shape charge with a large secondary shrapnel component so we can pierce the vehicle doors and have the best effect against flesh to ensure a kill."

Diane came by her cold-blooded passion honestly. She had grown up in the CIA's Clandestine Service, moving to technology protection programs associated with nonproliferation efforts before ultimately working the bin Laden target. She deserves much of the credit—some would say blame—for mentoring me and showing me how to develop nonlinear paths through the organization to obtain the desired outcome.

While in the Bin Laden Station, Diane's staff was well-supportive of her, although her leadership was not. Her key staff member was Bob, a former Army Ranger with a keen under-standing of the intelligence and a knack for plan development. Most important, he was forever skeptical of the intelligence being reported from the field. During the time when I was working in the Pentagon, it was Bob who was feeding our insatiable appetite for intelligence reporting.

Diane was also supported by Mary, a liaison officer from the DS&T. Mary was an exceptional officer who was able to reach back to her home organization for the technological needs of the Station. She shared Diane's passion and commitment to bringing bin Laden to justice.

By the end of May, I felt like I'd been named King of the Island of Misfit Toys, the most unlikely choice of all picked to lead a tech-nological insurgency to achieve what most people dubbed Mission Impossible.

As word of our challenge spread through the confines of our little community, we became the subject of an active betting pool. The house put us at long odds, with the smart money saying we would crash and burn spectacularly. But a part of me never believed in the smart money, not when the alternative was to believe in my team. With a five-foot-long Microsoft Project plan rolled up under my arm, flanked by as eclectic a crew as anyone could imagine or hope for, we set out to prove the smart money was wrong.

3: JUST DUCT TAPE AND BALING WIRE

MARK COOTER

COWBOYS, BANDITS, AND PIRATES
June 2000

"Contact me secure"—three words on an unsigned post-it note attached to Colonel Ed Boyle's business card were stuck on my STU-III secure phone.

As an Air Force major limping back to my Pentagon cube after a brutal and protracted fight with a kidney stone, I thought that finding the cryptic note was not the way I wanted my morning to start.

Colonel Boyle was the director of ISR[40] for the US Air Forces in Europe, headquartered at Ramstein Air Base in Germany. I knew him from Desert Storm and Allied Force. He had been one of the Intelligence Division commanders at the prestigious USAF Weapons School. Some refer to it as the Air Force version of Top Gun, but to Airmen it is much more than that.

---

40 Intelligence, Surveillance, and Reconnaissance, a collective term for the coordinated and integrated acquisition, processing, and provision of timely, accurate, relevant, coherent, and reliable information.

Colonel Boyle had always been all business with me in the past, and when he answered my call moments later he remained true to form. I was being pulled from my current dull assignment reviewing doctrine documents in the Pentagon. Hooray!

Then the other shoe dropped. None of the layers of lieutenant colonels and colonels above me were read into what we were going to do. If I needed something in the Pentagon, I was to report directly to Major General Glen Shaffer, Air Force director of ISR. That is a long way of saying that I'd be working for the senior intel guy in the entire Air Force.

My marching orders were simple: "Major General Shaffer has given you to me to work a special project; you are cleared from all other duties." The directions closed with a line straight out of a spy thriller: "Go see Snake."

I mentally traced lines up my current org chart. General Shaffer was my boss's boss's boss. Immediately intrigued, I almost ran through the Pentagon's wide corridors to get to Snake's office.

Air Force Colonel James "Snake" Clark was already something of a legend, certainly a bigger-than-life personality. A leading expert across the fields of mission planning, commercial imagery, UAV applications, and computer modeling and simulations, Snake developed Eagle Vision, the world's first truly deployable commercial imagery downlink system. The Air Force Predator program ran from Snake's office as well. However, many of my Air Staff colleagues held distain for Snake and would state many times, "He's true to his name . . . a snake." I just knew he got things done.

I knew Snake from my prior days in Predator, and although he caused me some angst during Operation Allied Force with some untested software, he always delivered outstanding successes. Besides, after spending my last ninety days plodding through a mind-numbing gauntlet of staff drudgery, the prospect of working with Snake held genuine promise.

My meeting with Snake lasted less than twenty minutes: jokes, jabs at intel, a pointed message about keeping this quiet, then "go do great things" and remember I "work for him." I wondered what the hell I'd been thrown into.

Luckily, also in the meeting was Lieutenant Colonel Ken Johns, Snake's deputy. Ken was a career Air Force intel officer like me and remained quiet throughout the meeting. I had heard that Ken was the action arm behind Snake's grand ideas, a guy who knew how to deftly maneuver in the Pentagon and make Snake's magic happen.

I gathered quickly that the current trick would have something to do with the magic of unmanned flight, even if the details were still sketchy. But my involvement started to make sense; just a few months ago I ended my time as a Predator Squadron operations officer. As the only Air Force intel officer to hold that position then and since, I would have bet that little novelty in my career path was done and over. It now looked like I was going to get to do it again. Awesome!

Ken had me join him in a meeting in the Joint Staff spaces. Sitting around a Pentagon conference table was Ken, flanked by Doc Cabayan, who served as our conduit to DOD. On Cabayan's right sat a slim-built, fair-haired Agency guy introduced as Alec Bierbauer. Given the other members of the team, he looked awfully young to be sitting at this table.

I shrugged. *Maybe he's an intern.*

As expected, the discussion centered on putting an unmanned aircraft into the air, which in and of itself was no big deal. But the devil really is in the details, and the list of arbitrary criteria being tossed around the table would have filled hell itself.

The aircraft needs to be able to read a license plate from twenty thousand feet and look smaller than a flock of birds on radar, all while providing forty hours flight time, and, oh, as a bonus, it has to operate over a target halfway around the world. The list sounded less like real-world capabilities and more like the makings of a good sci-fi novel.

My piece of the puzzle continued to come into focus. As an Air Force intelligence officer, I had a good handle on how to integrate Predator with the Agency's efforts. I was the only officer in the Pentagon who had been both a Predator Squadron's operations officer and sensor operator. I was one of a small cadre of people who had actual, hands-on experience flying what amounted to a big remote-controlled airplane through the skies over Kosovo and Bosnia.

But Predator was an ungainly beast that couldn't be allowed to wander more than a few hundred nautical miles from our base. This "wonder plane" they were describing, a near-invisible bird with Superman vision that could fly on the far side of the globe . . . well, that was fantasy.

As I listened to one criterion after another being thrown on the heap, a considerably less eloquent phrase began to predominate my quiet assessment: *This is crap*.

I quickly outlined my experience and a more realistic assessment of Predator's capabilities. That drew the reply "Well, General Atomics said it could do it."

In the back of my mind, I presumed the voice that made such a promise sounded a lot like the one that says "the check is in the mail" but that didn't seem the proper heading to take just yet. I took a deep breath and said, "Well, sir, yes, the aircraft might stay airborne for forty hours, but that's without cameras or SATCOM. And, yes, you can read a license plate of a car if you're parked on the ground." I told them I had been involved in the testing of radars against Predator and I could definitely say it was bigger than "a flock of geese."

I rolled down the list, feeling better to have the rebuttal off my chest. But I believed in their—now *our*—new mission. I had found my challenge, and it was time to roll up my sleeves and become the bridge between lofty expectation and the reality. Predator seemed to be the right platform for the job, which made this the right time to stress its capabilities.

It is one thing to tell yourself that almost any goal is possible if you bend the rules hard enough, but I was being asked to prove it before we would be allowed to deploy.

MR. VIDEO
July 2000

The term Air Force Base draws to mind an orderly array of immense hangars, control towers, and crisscrossed runways, a place where fighters and bombers proudly line a tarmac marked by windsocks and landing lights. El Mirage was something different.

Located just northwest of Victorville, California, El Mirage was little more than a flat spot in the desert. A stretch of cracked apron separated the lone runway from a line of prefab buildings and conex containers. Aside from the gray-white reflective rooftops, the entire place was a study in desert tan.

The GCS at El Mirage started life as a standard Air Force conex box. It was a huge upgrade from the GCS we had back at Indian Springs. The steel container was stuffed to the gills with ad hoc electronics; it was a command center that could be tossed onto an eighteen-wheeler or shoved in the back of a C-130. Back in the day, we'd broken two winches moving the old-style GCS out of Hungary and shut down a Bosnian airfield for hours on another deployment. By comparison, this one would be a hell of a lot easier to get to our deployed location, wherever that turned out to be. Save for an anaconda of cable running out of one end, the GCS that sat just a few dozen yards off the edge of the apron was, to outward appearance, just another steel box dropped in the sand.

Once inside, I stood behind the RV-style seats mounted in front of the dual pilot and sensor operator racks. Inside was Big and Ken Mitchell, the latter serving as the lead sensor operator from General Atomics. The two guys filled the RV seats like a couple of overage video gamers, joysticks in hand, facing an array of flatscreen TVs. A spiderweb of cables connected the wall of widgets, cords lashed to support frames with zip ties. The ad hoc nature of this endeavor was visible everywhere.

As the crow flies, the Nevada Test and Training Range was about 250 miles from El Mirage, a distance that an F-16 on a sprint could cover in about twenty minutes. For Predator, that trek would have been more on the order of a three-hour crawl. Thankfully, we didn't have to make that trek. Our job was to operate the aircraft over Nevada from El Mirage, California, while transmitting live video to Alec's team back at CIA headquarters in Langley, Virginia.

That last little stunt required something of a Rube Goldberg approach. We had the protocols, experience, and equipment for handing off control of an unmanned aircraft from one site to another. We verified that capacity before the aircraft left the ground. If we couldn't waggle the flaps on the runway, there wasn't much point in taking to the air.

But this trick play called for the Predator to lateral a video stream back to the ground station in El Mirage, which would then throw long—caroming that stream in real time off a satellite to come down at CIA headquarters. It only needed the phrase "Hail Mary" to round out the level of difficulty.

With bouncing signals off water towers and pilots not in planes, we were quickly creating a "who's on first" comedy routine that had become impossible to brief. We needed some new language. Sticking with the date that brought me to the prom, if stealing tech was working, I decided to steal some language as well. The air force Global Hawk program was my unwitting victim, although it lost nothing in the heist. What I came away with was a couple of, I hoped, clear terms to put all this into focus.

I grabbed a piece of paper, drawing a circle on the left side of the page that contained the letters MCE, Mission Control Element. That represented the collection of pilots, technicians, weather, and intelligence experts—largely Air Force personnel needed to fly the plane once it was in the air.

Inside the circle I drew a small rectangle, now pushing my artistic ability. Satisfied that I'd captured the essence of a conex container, I tagged the letters GCS to identify what amounted to a "cockpit in a box" in which the flight ops were conducted.

I drew a line across to the right side of the page, where for sheer diversity I went for a triangle, dubbed LRE for Launch and Recovery Element. This would be at an airfield, manned by a very small group of mechanics and a flight crew who were responsible for maintaining, prepping, and physically getting the plane into the air. The notion would be to maintain a discrete footprint, half a dozen self-sufficient "do-everything" types who could remain as invisible as possible. I wrinkled my nose; that wouldn't play well with Air Force tradition, which would have mandated two hundred Airmen and a golf course.

Once airborne, we'd throw a switch and, through some tech magic, control of the plane would leap from a direct link with the LRE right underneath it to way, way back to the MCE, ending up on the screen of a pilot sitting inside the cramped GCS. It was clear as hell on paper; now we'd see if any of that shit would actually work in the real world.

Given the urgency of this experiment, one might think that we had an arsenal of high-tech military hardware to pull this off, or a sleek bit of James Bond spy craft developed by Q Branch. What we had was Mr. Video.

A roving relay-for-hire, Mr. Video was a commercial service that provided in-field uplinks to cover things like local wildfires or golf tournaments. It was fundamentally one guy and a panel truck chock-full of broadcast gear. Absent anything remotely like an alternative, we found Mr. Video in the yellow pages.

The job was "eyes off"—Mr. Video had no idea what he was transmitting. He would connect his truck to a cable, establish the link, then wait outside his truck until the job, whatever that job might be, was completed. He was actually a little miffed that we only needed him to transmit a single video stream.

"That's it?" he grumbled. "Pffft, even golf is more exciting than this." If he only knew.

I glanced out the door and confirmed that he was on station, his butt parked in a lawn chair with no clue that he was orchestrating a test that would demonstrate we could pump video of sensitive targets from anywhere in the world to the CIA's inner sanctum. That little bit of improv likely bent or broke any number of rules, but we weren't under orders to be model citizens; we were told to get the job done.

We stood inside the ice-cold GCS, watching video of bleak, rugged terrain scroll by beneath our point of view. The view was not unlike our ultimate goal environment: jagged ridgelines carved by meandering arroyos and wide expanses of nothing at all.

I looked across to my right where Alec stood as I did, staring intently at the mosaic of screens. What exchanged between us was a wordless mix of a shrug and a nod. The success criteria going into this experiment was no more specific than what engineers called the 50/50 rule—either it's gonna work or it ain't. I watched the arid tones of Corn Creek Springs, which represented our sample terrain loosely simulating the desert near the foot of the Hindu Kush mountains of Afghanistan, drift silently across the screen. We had no doubt that we'd made it work.

Alec picked up the STU-III, the encrypted telephone that gave him a secure line to CIA headquarters. The call was brief. Alec started with "yes sir" then repeated himself several times before ending with "roger that." Success calls are usually short; the other kind take a lot longer. Alec placed the black handset on the cradle and said with a faint grin, "We're good to go."

Milestones are supposed to bring at least a moment of rest, a chance to pause and appreciate the accomplishment. At that moment the only thing going through my mind was the start of a new and far more demanding clock and the million things that had to get done.

The first, I realized, was that I had to thank Mr. Video and get him off the airfield.

WELCOME TO GERMANY
August 2000

There were moments when the progression seemed almost surreal. Over the span of just a couple months, I had been whisked from an endless, fluorescent-lit purgatory of Pentagon administrivia to sweltering in a blighted, dusty wasteland. Tonight, in a subsequent act of magic, I was driving through the lush green of Germany at Ramstein Air Base, wrapping up the process of assisting the heist of a ten-meter parabolic satellite antenna from the Air Force. Given a deployed diameter of over thirty feet and a couple added tons of motors and control systems, you are talking one Big-Ass Dish, and so it was known. The BAD was so big, it turned out, that precious few existed and nobody that had one was willing to let it go.

Despite the amorphous comfort of top cover that came with my "get it done" directive, being in Germany reminded me that the response "I was just following orders" didn't always cut it. But the mission parameters were clear, and I needed the dish, more so than my fellow Airmen back at Langley Air Force Base. Still, as I slid deeper into an unaccustomed trend of arguably felonious behavior, not getting caught moved steadily up my priority list.

Ramstein is the type of air base you see in the movies—a bustling city of its own built around a pair of ten-thousand-foot runways. Ramstein had seen service as a NATO command center and had been home to many diverse missions and units over the years. Today, the host unit was the 86th Airlift Wing, which operated Ramstein's expansive tarmac and ever-changing collection of airlifters, tankers, and special-mission aircraft. In the midst of all that was us.

We took over a small, postage-stamp-sized compound near the large Air Mobility Command ramp inside the Ramstein perimeter. Our tiny domain was surrounded by a chain-link fence, with an armed sentry stationed at our front gate. The cover story for our existence was only marginally more water-resistant than the sixteen-by-thirty-two-foot GP medium military tents that served as our offices on the compound.

In addition to serving as all our intelligence analysts, Airmen from Colonel Boyle's staff and the 32nd Air Intelligence Squadron's team, led by Ginger, conducted most of the bed-down tasks such as reserving billeting, setting up tents, you name it. When I needed something, no matter what, I called on Ginger or Colonel Boyle's executive officer, Air Force Lieutenant Brian "Rio" DeGennaro and it was done, no questions.

One night during a hard rain, as we continued to prep and improve the facilities and close possible failure points, I noticed that our fiber transceiver was sitting in a puddle of rainwater.

I turned to Cliffy, pointed at the device, and asked, "Hey, is this a problem?"

Cliffy Gross was the embodiment of the can-do senior NCO. No problem was too hard. Experience had shown that, once fortified with a carton of smokes and a two-liter bottle of Diet Coke, Cliffy could fix or build just about anything. If the day ever came that he got beamed up to the Starship Enterprise, Cliffy would put Scotty out of a job.

At the moment, Cliffy was crammed behind a rack of mission comms gear. He craned his neck in my direction, squinted at the transceiver, and asked, "Is it working?"

I squinted at the gizmo. The lights were blinking. With a shrug I replied, "I guess so."

"Then it's not a problem. Leave it alone." His attention reverted back to the comms rack.

I stood for a long moment, slowly shaking my head before returning to my prior task. I'd come to trust that if Cliffy said it was good, it was good.

Security was an ever-present challenge, in ways one would rarely expect. The odd thing about guard duty is that all too often a junior Airman is stuck at a gate having no idea what, or who, he was protecting inside. As the new kids on the block, many of us were a bunch of unfamiliar faces and not always quite dialed-in on the Ramstein security protocols. Every now and then that would translate into unintentional friction between the protectors and protectees.

I was nose-down in the GCS one evening when it was made urgently clear that my attention was needed outside. I came out at a trot to find a disgruntled Gunny staring down the barrel of a flustered Airman's M16. The kid was mechanically repeating a "none shall pass" ultimatum, despite a crumbling tone of confidence as Gunny took another step forward. My worry wasn't that the Airman would actually shoot, but that Gunny would wrap the rifle around the poor guy's neck. I quickly calmed the situation, and the Airman survived.

Not all of the miscues were resolved at the gate with a timely intervention. One night I got a call to come spring Cliffy from jail, having been locked up by an overzealous guard who perceived Cliffy as having "crossed the red line" in the process of doing his job. That one could have gone south quickly; once paperwork is generated on an incident, it takes on a degree of self-expanding inertia, no matter how absurd. Luckily, a local chief master sergeant beat me to it, and Cliffy was released.

Having been shipped in from El Mirage, the GCS sat on a checkerboard of concrete slabs just a short walk from the gate. It was painted a tan-brown-black camo, nondescript until Gunny plastered a bright Big Safari stencil on the side. So much for secrecy.

A bit of convenient fiction was offered to explain our presence. The entire assembly was passed off as part of the Air Force DCGS program.[41] Itself a collection of regionally aligned, globally networked sites, the DCGS was a natural cover for a bunch of guys with a truckload of satellite gear. On paper we were just an offshoot of DCGS or Snake's Eagle Vision program, and no one was the wiser.

Still, as the new guys in town we had a number of challenges fitting in with the standards of acceptable behavior and decorum. Our anemic veneer did little to explain the envelope of secrecy that quickly wrapped itself around our little enclave.

For all of the high-tech aspects of our ad hoc outpost, some aspects were decidedly suboptimal. Living on a concrete pad in the middle of an airfield meant that no plumbing was anywhere in the area, so porta-potties were an unsavory necessity. That would hardly draw a blink in most contexts, but with all of our gear crammed elbow-to-elbow within the confines of our fence, the shitter was but a few steps from the GCS door. That made for

---

41 Distributed Common Ground System. Formally designated the AN/GSQ-272 SENTINEL weapon system, the DCGS is the Air Force's primary intelligence, surveillance, and reconnaissance collection, processing, exploitation, analysis, and dissemination system.

some pungent work environments, aggravated by that fact that the cleaner contracted to pump out the tank was at best erratic in terms of schedule. The extent of the problem was made abundantly clear when Ginger burst into the ops tent in a snarl of Kentucky twang, announcing that there was "shit up to her ass" and that something had damn well better be done to fix it.

Ginger was a tall, athletic Air Force intel captain who had a fire like nobody I had ever known. Unmatched in her passion, commitment, and patriotism, Ginger was not a force to be trifled with. Serving as the Air Force 32nd Intelligence Squadron assistant ops officer, Ginger had been a key acquisition when Colonel Boyle set out to cherry-pick the best of the best for this team. Her down-home drawl came out the strongest when she got fired up; when that happened here, the word *shit* was pronounced as it was intended in the south—with two syllables. Anybody with a brain took the emergence of drawl as a sign to listen up or get the hell out of the way.

Not all of the problems were quite so linear. Because no TV shows or soccer matches were beamed from Afghanistan, only one satellite could provide us the necessary communication links to operate our Predator. Our normal PPSL[42] couldn't handle the low-look angle, gain limitations,[43] and low-power requirement to maintain our link. Even with the Big-Ass Dish in place, we determined that a second smaller dish would be needed to carry the command link because the available satellite transponder operated outside the frequency range of the PPSL.

---

42 Predator Primary Satellite Link, the device we used to conduct Beyond Line-of-Sight satellite communications.

43 In the most basic of layman's terms, "gain" is a numeric representation of how well you can transmit or receive a radio signal.

As Big Safari's resident genius, Albert had worked out a back-room deal for a suitable dish found in Italy. It wasn't everything we wanted, but it was cheap and readily available. Secrecy prohibited having the sellers deliver it to Ramstein. Instead, the sellers were directed to haul it to a stop on the autobahn, where some of Paul's Airmen took possession and drove it the rest of the way on a flat-bed.

Strung between all these parts sat a couple of generators, along with an empty square of ground just big enough to shoe-horn our recently purloined Big-Ass Dish alongside the LBIT—Low Budget Italian Terminal.

Air Force Captain Paul Welch was a part of that acquisition. A technical comms expert from the 1st Combat Communications Squadron, Paul had cut his teeth creating networks in some of the world's most inhospitable locations. Another make-things-happen kind of guy, Paul seemed impervious to the challenges raised by rough terrain or lousy weather. To have Paul and Cliffy together on one team was a miracle, yet another testament to the mastery of Colonel Boyle leading the Air Force equivalent of a fantasy-football dream team.

4: ALL PRICES ARE NEGOTIABLE

ALEC BIERBAUER

IGNORE THE MAN BEHIND THE CURTAIN
August–December 2000

At moments, I could look across my to-do list and feel like I was making measurable progress. But whatever sense of accomplishment that might have warmed my heart was just as quickly torn away when I flipped to page two. The top item read, "Construct a remote-controlled, top secret airfield on the far side of the planet."

Although sounding like a page from some super-villain's diary, a forward air base was an obvious critical requirement. Our launch site had been selected as being close enough to Afghanistan to spend the majority of our flight time over the country, yet far enough away to maintain a secure facility with the requisite infrastructure. After limited debate, the chief of station had thrown his unqualified support behind the program and positioned the pitch to the host government.

The final discussion required sending a representative from headquarters, and I was happy to see it through. It was one of those whirlwind trips—two days' notice to secure a visa, then a two-day trip to participate in a one-hour meeting to hammer out the fine print. I was back out the next day.

Like so many other parts of the program, the fragile nature of all this hung in the balance of travel problems. The meeting was set, but my United Airlines flight landed late in Frankfurt, and I missed my connection on Lufthansa. The next available flight would be in twenty-three hours, about fifteen hours too late for my meeting.

I went to the local airline counter and asked if I could wrangle my way onto a flight just two hours later. The agent said it didn't look good, but we could ask his boss when he arrived. Said boss strolled in and explained to me that my ticket could not be honored. With the fate of the program possibly hanging in the balance, I asked what a new ticket would cost, and after a long moment of sizing me up he said with a shrug, "Maybeee . . . one thousand dollars?"

"Deal," I grumbled and slid my credit card across the counter.

He wrinkled his nose and added, "Cash."

It was the way things worked in the third world; the price for something is a matter of what the market will bear. Most, if not all, of the ticket price would go no further than the manager's pocket. Frankly, I didn't care; in the grand scheme of things, I was confident it was a fair use of taxpayer dollars.

Reaching my destination at last, I was met at the airport by Fred, the CIA chief of station. It was our good fortune that Fred had a solid rapport with Diane. She schooled me in the nuances of DO[44] etiquette and laid the groundwork for my meeting with Fred, who otherwise would have likely had me for a midday snack.

Instead, thanks to Diane, by the time I sat down with Fred we had a good handle on the political and personal landscape. Fred was no fan of the seventh floor, the CIA's executive suite, and would sign up for this program just for the sport of it. If I were a betting man, I'd say that Diane preceded our meeting with an off-the-record chat with Fred as well.

It was well known that with Fred everything was a negotiation. He was seen within the Agency as an unorthodox, perhaps even contentious, figure. Being aggressive and outspoken might be mission-critical for a station chief, but it didn't win Fred a lot of friends within the CTC.

But Fred was also a pilot, and like any pilot he was intrigued, if perhaps a bit uncertain, about the idea of an unmanned aircraft. He had built a strong relationship with the US ambassador as well as the host government.

Fred would support a pitch, but he wanted a manned aircraft at his Station as well, ostensibly as a means to score himself some stick

---

44 Directorate of Operations. The CIA had four Directorates: Operations, Intelligence, Science and Technology, and Support. The DO was also known as the Clandestine Service and was responsible primarily for recruiting spies and stealing secrets from foreign governments or organizations to further the defense of the United States. The DI performed the analysis and produced intelligence products for US government customers, primarily the White House, Congress, and the other departments such as State and Defense. The DS&T provided the technology to support the DO. Support kept things running smoothly through facilities, logistics, personnel, etc. All were dependent on each other. The DO ran very much on the basis of relationships, and Diane was able to leverage her lengthy career's worth of relationships to open some critical doors where I, as the new guy trying to go it alone, would have met much greater resistance.

time in the process. Another item went into my checklist: "Find a plane for Fred."

Satisfied with his wrangling, Fred hustled me directly to our meeting with the government. As it turned out, he had done a fine job of prepping the battlefield, and the meeting proved to be largely a formality. Not ones to miss a chance for quid pro quo, they did have some terms of their own, to include our support for having a greatly despised local militant group formally designated as a terrorist organization. They also wanted our assurance that we could be quiet and discrete with our operations. The politicians sitting across the table were not interested in stirring things up with the Taliban or al-Qaeda. Allies are allies, but nothing comes for free.

It struck me as the meeting went on that having someone with Fred's skill handling the situation was a remarkable break. A guy with his résumé would be eligible for consideration at the most prestigious stations around the world.

This was not one of those places. Getting stuck in the ass-end of the world was a fate reserved for a mid- to senior-grade officer with considerably less experience and standing in the community. But Fred was married to a junior officer who was well suited for one of the local staff positions, and Fred was able to wrangle the top spot in order to stay with her in a tandem assignment. He was the old-school case officer who, as chief of station, "owned" the country from the lead-for-intelligence perspective and was only accountable to the ambassador for the majority of his actions.

A seldom-followed saying among the Clandestine Service was to never "case officer a case officer." Much the way we identify hot buttons with our targets, the same is done on a routine basis to each other, and even to our leadership. In Fred's case, he had a firm grasp of the country, the intelligence structure, and the US ambassador. In my case, I had the ability to offer a support aircraft on standby to aide with the logistics and manpower requirements for our Predator deployment. Further, we were offering Fred about as controversial of a program as there was at the time, with a real ability for him to be on the front page of the daily updates to the DCI[45] and also to the president via the President's Daily Brief.

Little did the Clandestine Service realize when it parked Fred in a sleepy Central Asian country that his little oasis would become our staging point and a critical ally in the actions against al-Qaeda in Afghanistan. Fred's support would remain key throughout the selection and development of our infrastructure as well as in maintaining the clandestine status of our local operations.

Step one in that process was selecting an airfield. According to Air Force standards, Predator required five thousand feet of high-quality paved runway to operate safely. Not many airfields of that quality were in the region, and far fewer could be discretely used by our program.

NIMA satellite imagery proved invaluable in vetting candidate sites. One was selected, and the locals gave us their blessings with the contingency that we address their internal militant problem. Fred accepted that challenge, and I committed to carry the message back to HQ as well as to plant the seed with Roger and Dick at the

---

45 Director of Central Intelligence. At the time, DCI George Tenet sat at the top of the intelligence food chain. He held a daily meeting in his conference room, and there was competition to be on his agenda. People coveted face time or the chance to push favored topics in front of the DCI. Predator was a near-daily topic, given the complexities of the program and the potential blowback if something went astray. Though only cautiously optimistic about our chance for success, the DCI gave us considerable leeway to design and build the program as we saw fit, while keeping a watchful eye.

NSC.[46] Ultimately, it was an NSC and State Department issue, and we would need some support for getting it done quickly.

The next step for our program was conducting a survey, putting up a fence, building a hangar, and building out an airspace-usage plan. The prospect of operating a clandestine airfield is daunting; history tells us the idea is a recipe for failure. Try to imagine how you would go about concealing an aircraft, not just on the ground but also in the air as it takes off and lands. Nothing short of fencing in a hundred or so square miles of land would keep a plane in the open sky out of sight.

Add to that the necessity to interact with air traffic control and the untold dozens of people who are in that knowledge loop. The gravy on top included the ant line of food, medical, fuel, billeting, sewage, and other daily services needed to support a small, isolated community.

We needed to be clandestine for numerous reasons. Topping that list was the nature of the program's intent, that the CIA was executing it, that the White House and NSC had endorsed it, and that the host government was supporting it with the express contingency that it not be identified as complicit. I can't say how many people thought we could actually keep operations out of sight versus those who felt it was just important to try. I was on the side of it being really important to try, especially since our new landlords could shut it down as fast as they approved it.

---

46 The National Security Council. Dick Clarke served as the counterterrorism coordinator, and his deputy, Roger Cressey, focused on the counterterror problem set—specifically bin Laden. Dick and Roger had some organizational leverage with the CIA but weren't afraid to use brute strength to pursue an agenda. As the architects for the original memo signed by National Security Advisor Sandy Berger, it was a passion for Dick and Roger to shove the Agency out of its comfort zone. I believed in their passion, which left me at times to have dialogues with them in the background to further our collective interests.

We put a ceiling of not more than ten US persons on the ground at the airfield. We also made it mandatory to prohibit US flagged aircraft from flying to the site. All ground vehicles in support had to be locally purchased. All infrastructure improvements would be done with local contractors, with oversight provided by the US presence. Americans on-site were confined to the camp or flown out directly to the capitol on Fred's new airplane.

The Air Force was completely supportive of trying to operate within these constraints, but it drew the line at deploying any Air Force bodies. We were able to use Agency personnel and put a contract in place with General Atomics to get their best guys forward to be the launch and recovery element for this new approach to doing business.

One of the biggest challenges was that we wanted all Predator flights to and from the airfield to be done under cover of darkness. This would be the best way to avoid having people in the surrounding farms and communities glimpse this odd-looking plane and figure out that something sketchy was going on.

When we began to occupy the site in the summer of 2000, we did a fairly decent job of keeping a lid on our activities. The heavy lifting would be provided by Air Force C-17s. The major runs were supplemented by short flights with some smaller aircraft like King Air and Dash 8. Ground movements took place in local vehicles.

As far as the locals were concerned, a new tenant was in town. But with a host-nation security force and limited direct contact, the airfield looked no different than any other local government or commercial operation in the area.

5: SHOWTIME

MARK COOTER

MURPHY WAS AN OPTIMIST
August–September 2000

Whereas the GCS and other control parts only had to make a comparatively short hop, two Predator aircraft had to make the long haul from El Mirage to Alec's newly established base, a distance of over seven thousand miles. *Two* is an important number in the world of aircraft; redundancy is everything in the military, where things have a way of breaking down—or blowing up. Two is one, the axiom says, and one is none.

Therefore, the shipping logistics were top-notch. Nobody wanted to put the one-is-none axiom to the test on day one. Each aircraft had been disassembled and packed into what we called a coffin—by all outward appearance a thirty-foot Pelican case on wheels. Though containing all the aircraft and the tools needed to assemble them, the shipment did not include a GCS, at least not a normal one. What it had instead was the highly miniaturized version that GA developed to run from inside a nuclear submarine. While lean and without frills, it would allow the local pilots to run system checks as well as handle take-offs and landings.

Once in the region, the coffins and support equipment were taken off the Air Force C-17 Globemaster under the watchful eye of Alec's people. The cargo was then cross-loaded into a nondescript but decidedly civilian Lockheed L-100 for the last leg to the secret airfield. That swap was driven by more than the prohibition on US-flagged aircraft. The out-of-the-way airfield would not hold up to the weight of the heavier C-17. Yeah, a really great pilot might be able to set one down, survive, and take off again, but nobody else would be taking off again. The prop-engine L-100 could make it in and not tear up the runway.

Of course, the ever-present Murphy's Law[47] could not be ignored, and unexpected hurdles routinely cropped up. Case in point was the night we realized that the pallets that easily fit in the belly of a massive C-17 couldn't be shoe-horned into the smaller L-100. Repalatizing cargo was a bitch on a good day, much more at a distant airfield in the dead of night. Channeling my inner cowboy, I knew we needed to head that problem off at the pass now.

The ad hoc posse that ran that problem down included Jim Klein,[48] the General Atomics site lead, and Tom Coffield from Big Safari. Together we repalletized the entire forward package right there on the tarmac. To fully understand this decision, close your eyes and imagine Santa and his elves unpacking and repacking the entire sleigh in the middle of a busy runway on Christmas

---

47 Murphy's Law states, "If something can go wrong, it will." Corollaries of Murphy's Law suggest that absent any other contributing factors, this failure will take place at the worst possible time.

48 Jim was a retired Navy senior chief petty officer. If I needed something, he'd get it done. If we needed something bought, he'd buy it, putting him at risk with his GA comptroller. Luckily, his GA leadership, especially Brad Clark, fully supported our efforts. In 2001, I asked him to act as my first sergeant and look after our team. Without blinking, he accepted and performed phenomenally. Jim embodied the "just do it" essence of our team.

Eve. Then deck the elves out in Air Force battle dress uniforms, and you'll have a pretty good mental image. None of the other reindeer, all desperately trying to fly out on that runway, were at all amused.

Once delivered to their final destination, the coffins would be opened, and the aircraft and support systems assembled. This was not a trivial task, requiring a few days to complete. Ground checks followed, testing every flap, actuator, and blinking light. Given the scarce availability and cost of the aircraft, much less the million-dollar surveillance ball hanging under its nose, there was no such thing as a cheap fender bender.

I had received confirmation that all of the ground checks came back green, leaving only the FCF between us and our first live mission. The Functional Check Flight is just as it sounds; putting the aircraft into the sky and doing a lap around the track. The process would be familiar to oval-track racing fans—basically, four left turns connected by an equal number of long straightaways. An FCF was mandatory Air Force protocol after any disassembly that went beyond basic maintenance and certainly applied when an aircraft had been stripped damn near down to the bolts.

Things had been going well, too well for Murphy, who once again exerted his invisible influence on the simple act of an FCF. The problem stemmed from the CIA's insistence that the flight be done at night. This was not the normal way to conduct an FCF. We had gone a couple rounds on that point, but the obvious additional difficulty of flying in the dark lost out to the demands of secrecy. The Agency argued, better it should crash in the dark then succeed in the day and be seen by the enemy. *I'll end up reminding Alec of that one day,* I thought to myself as I hung up the phone.

I was in the Ramstein GCS when Predator 1 spun up its engine on a dark, distant runway. Alec was next door in one of the tents along with Mike,[49] the Agency comms guy.

They say no battle plan survives the first contact with the enemy. In our case the first flight plan didn't get us much past revving the motor. Despite all of our cross-checks and contingency mapping, we suddenly found ourselves looking at something new: night. We couldn't see shit through the nose camera at night.

A small lens set in the tip of the plane, the nose camera existed to provide a constant forward view to the pilot. But the device was purely optical, and we were launching in pitch black. At that point, the nose-cam could see no better than the human eye. Looking forward into the darkness, our plane was as blind as a bat.

Under Predator's chin, however, hung the all-seeing eye. The huge polished lens of the Wescam Versatron 14TS could suck in faint traces of both starlight and infrared to peer through the night. Normal flight protocols call for the ball to be caged—that is, rotated to a forward-facing position and locked into place below five hundred feet for takeoff and landing. This practice was developed to give another video for the pilot to use in case the nose camera failed. Once in flight, the ball would be uncaged, free to swing around on its own to view whatever the crew needed to see.

The plane plodded down the runway, and with an utter absence of ceremony, climbed into the night sky. Upon reaching test altitude, Predator began to plod around the circuit like a NASCAR

---

49 Mike was a retired CIA case officer who signed on to assist as the forward liaison and communications officer at Ramstein. Because this was a new program, stood up with no notice, we faced a shortage of qualified personnel to fill critical roles. Mike was a perfect fit. He kept Alec and the team well informed and looked out for the Agency's interests.

ᴄontender limping around the track under a yellow flag.[50] At each corner the camera would swing left, and the pilot would steer likewise to align with the adjusted view. Four textbook turns later, we were done. The camera locked in on the approach end of the runway, and the pilot settled into final approach. The only thing left to do was set it down.

I knew Alec would be in one of the two tents next door, on a secure phone with his bosses in Virginia as they all watched the same video stream. Success was beginning to feel a bit routine, diminishing perhaps that the test we had just flown once again broke new ground for remote-piloted flight. I watched the aircraft sink slowly toward the asphalt, just a couple meters off the tarmac as it crossed over the near end of the runway.

Suddenly the image on the screen gave a sickening lurch, whiplashing left as though in a tumble. The pilot reflexively yanked his stick to starboard to counter the port roll. The screen went black.

Barely a second ticked by before the phone rang. Alec was both confused and concerned. "We lost feed there at the end. What happened?"

My brain was trying to make sense of the spinning jumble that had just flashed across the screen. I couldn't venture a guess, so I clung to what little was clear. "They had a hard landing."

"Ah shit," Alec groused. "I just sent word to Rich and the director that the test ran perfectly."

I palmed my face, head shaking. Rule number thirty-seven: never give a report until you have photographic evidence and the analysis to back it up. I didn't envy Alec having to amend that little synopsis, my fleeting shred of pity dialing back the tone with which I might have otherwise replied.

---

50 To be fair, NASCAR drivers can break two hundred miles per hour on the track; with a good tailwind, we were lucky to hit half that.

"Well, you better go unscrew that one." I said, then hung up, focusing my full attention on unraveling the disaster.

Over the next hour, my team reviewed the footage and collected data from the guys downrange. The truth, once understood, taught us yet another new lesson about unmanned flight.

In its target mode, the camera ball stays on a designated point on the ground. That function is invaluable when you need to keep an eye on a given point while maneuvering the plane through the sky. But in preparing to land, the crew inadvertently locked that gaze on the near end of the runway and left it in that mode versus changing to position mode. As the aircraft flew over that point the ball dutifully spun around to keep watch on that specific point. While the plane was on a smooth, straight path the camera snapped a one-eighty. This disoriented the pilot, causing him to abruptly move the flight controls, resulting in the loss of the aircraft.

A pilot with his ass in the aircraft would have known there was no gut-wrenching lurch, but a pilot on the ground has only his eyes to rely on. Seeing his view corkscrew in one direction, his natural move was to counter, and counter hard. In doing so he flipped a perfect landing into a multimillion-dollar tangle of aluminum and carbon fiber skin that littered the runway.

I walked out of the GCS to see a darkening sky overhead. The night that had swallowed our launch site several hours ago was just now washing over Germany. Off to the west, the last fiery crescent of sun sat on the horizon like a burning wreck at the end of a long runway. Perfect.

A tent flap swung open, and Alec walked out, hands on his hips. He looked about two years older than when I'd last seen him earlier today.

"How'd it go?" I asked.

"You don't wanna know," Alec said with a tired grimace. "I told him there was good news and there was bad news. The bad news is that we crashed the plane, the good news is that the pilot survived."

I looked at Alec for a long moment, trying to figure out if that was just his sense of humor or whether that was how reporting was done inside the CIA. I wasn't nearly as sanguine as to how this bit of news would be received up my own chain of command. The operation was barely out of the box and in a world where "one is none," we were already down to one aircraft.

MISSION 1
September 7, 2000

Having crashed half our fleet doing a simple lap around the runway, the mounting pressure of our now "one-and-out" status hung heavy on my mind as we gathered in the GCS on September 7. This was Mission 1, our first actual launch into enemy airspace.

I looked around the GCS where the team stood or sat at their assigned stations. Big was in the pilot seat. He had more stick time flying a Predator than just about anybody on the team. Gunny sat next to him, eyes fixed on the sensor screens. Steve "Cruiser" De La Cruz and Swanson were on deck to take over pilot duties later, and Ken Mitchell, Steve H., and Todd G. were stacked behind Gunny.

Beyond all of the tests, all of the handwringing, we still faced a number of unknowns, a number of questions answered with the phrase "it oughta . . ." We grappled with issues like the quality of our satellite feed. The models and simulations suggested we would have decent satellite coverage at the airfield, but reality indicated it was spotty at best. We were assured by Albert that it would get better the farther south we went. He said repeatedly that "it would be good over the primary target areas of Kandahar."

While we were reasonably comfortable that our host's military would cooperate and not blow our plane out of the sky, we had no idea if Afghan radars, on the ground or in the air, would pick up a gangly silhouette trudging across the sky or that the Afghans would know what to do if they picked it up on radar. We were looking at undefined variables on target data, on weather; in fact, damn near everything past the end of the runway was a huge unknown. The only way to know for sure was to go there.

In the GCS, systems and backup systems were humming to the extent that we had them. A couple thousand miles away, our plane sat on a runway outside the LRE. The little triangle I sketched back at El Mirage was now real, at an isolated runway in a remote corner of a nation few Americans could locate on a map. Our Mr. Video trick had been a demo; this was the real deal. With machine precision, we went through our ground checks.

The LRE team handed us control to confirm that we could operate the bird from nose to tail—every flight surface, the engine, the brakes, all the way down to the complex million-dollar sensor ball. We ran the checklist by the numbers before relinquishing control back to the forward-team guys to put it in the air.

We watched through the IR sensor on the ball as the aircraft rumbled down the dark runway, watched as the horizon line dipped down out of view and the screen ahead filled with nothing but night sky. Our forward launch team, consisting of Gil G., "JDOG," Mike Johnson, and yet another Paul,[51] had done a remarkable job

---

51 Volunteers from General Atomics, these four guys represented the first contractor technical team deployed to the forward location for launch and recovery. The set-up was as bare bones as you could get. Everyone needed to know multiple jobs and needed to stay healthy. When you have only four players on your football team, everybody stays on the field, switching between offense, defense, and special teams at the drop of a hat.

under extremely austere conditions, surviving untold stress to give us the perfect start to this first mission.

Once the aircraft was safely up to about ten thousand feet and everything appeared normal, we would work the handoff to the pilots in the GCS. The launch team and mission control team ran our hand-off checklist and control passed over with an unceremonious "It's all yours."

The plane climbed to altitude and began to follow an exhaustively calculated flight plan into Afghanistan. I glanced at the clock. It would be a while before the aircraft crossed into Afghan airspace. Now was the last chance to detect any final hidden flaw before driving America's most secret aircraft over the heads of our enemy.

The last checklist was binary: a simple go or no-go.

"SATCOM?" I asked.

Cliffy responded, "SATCOM go."

"Weather?"

"Weather go."

"Targets?"

"Targeting go."

"Threats?"

"Threat is go"

I checked the boxes down the page, tension rising with every "go" that Ginger passed.

A final check remained, and I pinged Alec on the phone. "Everything is go here."

Alec was tag-teaming mission management for this first flight with Hal in the GRC and was equally subdued in his reply, the weight of the moment not lost. "Roger says comms are clear, nothing is spun up that looks like a hazard. Time to make history." He paused for a moment, then added "Oh, and Mark . . ."

Here it comes. "Yeah, Alec?"

"Don't fuck it up."

We'll do our best, I thought as I sat the clipboard on the desk. Save for the hum of electronics, the room was dead silent when I said to Big, "Pilot, you are cleared to cross into Afghanistan." Somewhere in the darkness far, far away, unseen and unheard in the night, our plane went to war.

As dramatic as that might sound, the first obstacle was the long plod down to Kandahar in southern Afghanistan. The term *roughly* was not a trivial; with an aircraft as light and sprawling as Predator, a stiff headwind could bring our groundspeed to a veritable standstill, and we would need to tack east or west for a bit much the same way a sailboat would. That could result in an abort heading outbound, but it could spell disaster on the way home.

USCENTCOM had tasked only one guy to watch Afghanistan military orders of battle, which is a fancy military term for what type and how many aircraft, vehicles, radars, and troops it had. Therefore, the orders of battle were very suspect and outdated. We worked on the assumption of Russian-based central C2 systems that put three primary early warning radar sites and one SA-3 missile battery in a west-to-east line that centered on Mazar-i-Sharif. The early warning radars could see out to long ranges and were overlapped in coverage to create a picket-line of detection and intercept capability.

Our plan included the use of what we called ops clocks, an extensive exercise of threat mapping based on observations collected over time. Threat data was plotted among numerous pie-shaped wedges around the clock.

Based on extensive analysis of the patterns of up and down times of the radars, spider routes, linear pathways connected by a series of points, were developed to exploit terrain masking or time-sensitive gaps in radar operations. In some places, it was a matter of slipping through a narrow gap or dashing across the yard when the lights went out. For every flight path, we had contingency flex plans we could invoke at short notice if a hazard presented itself.

We watched as the aircraft snaked its way down through the Hindu Kush mountains toward Kandahar. Some five hundred miles of bleak, jagged peaks as high as up above twenty-five thousand feet, the very name of the range spoke of its deadly heritage, argued by historians to trace to the ancient Persian phrase for "Hindu Killer."

Our control link bitched "beep, beep, beep" incessantly, warning tones that advised us minute by minute that our satellite footprint was far from optimized. The pilot's additional duty became to constantly push the reset button to shut it up. I kept reminding myself, *Albert said it would get better.*

But it didn't get better. The reset was so bad, the pilot had to set our lost link start point ahead of us to prevent the aircraft from turning home each time we lost the link—yet another risk we took in order to give us a chance at mission success.

While misleading by its name, lost link was a safety mechanism to use GPS to send the Predator on a predetermined path back to the home airfield in the event satellite or line-of-sight communications were lost with the aircraft. The standard protocol for the Air Force is to put that lost-link orbit in a safe area within line-of-sight communications range to the airfield. The aircraft protocol would then be to circle there until the communications link is regained or the aircraft runs out of fuel in an area with minimal risk to people or facilities on the ground.

For our purposes, having the aircraft return to the LRE would create problems with air traffic control in Uzbek airspace and degrade this critical ally's trust and confidence in our operations. The established protocol with our host government is that we would only operate on the friendly side of the border, a relative term at best, during hours of darkness and low air-traffic volumes.

A lost-link protocol was designed to navigate an aircraft back to its launch point if communication with the plane was broken. That wasn't some hypothetical; given the weak satellite signal over the LRE, we would routinely lose contact with Predator as it headed into Afghanistan. This is where we gamed the system. As we moved into our known black-out areas, the pilot would continually push the "return to here" pin on the map to a point ahead of us, further along our flight path where signal would be better. Basically, if the plane dropped the call, it would fly ahead to the address of a known pay phone. This bit of fudging kept us from incessantly making U-turns as the plane lost and regained satellite contact.

Once we crossed the border, the pilot had to build lengthy orbits over the Hindu Kush mountains into the lost-link profile to avoid returning home during daylight hours. Although not an ideal situation, we could avoid problems with our landlord. The unpopular term suggested an expectation of failure, but the simple reality was that absent a lost-link emergency mission capability, we could end up with a lost airplane.

On the bright side, rugged mountains worked in both directions. The Afghans had little-to-no radar coverage through the peaks and valleys of one of earth's most desolate terrain.

It felt like an eternity had passed when we cleared the southernmost ridgelines and slid out over the flat expanse of dirt that led to Kandahar. In reality it had been several long hours. Leaving a comparable time for travel home, that left us fuel for about six hours on-site to see what we could see.

Then again, "hours of fuel in the tank" isn't the same as "hours of unmolested time to wander around in hostile airspace." Boasting a full military airfield, Kandahar was defended by Soviet-era SA-3 surface-to-air missiles as well as MiG-21 jet fighters. An eastern bloc workhorse that dated back to the Vietnam war, the supersonic MiG-21 could break 1,300 miles per hour in a full-on sprint. That was about ten times what the Predator could achieve in a nose-down, full-throttle fall from the sky. We had to rely on keeping a low profile because we sure as hell weren't going to outrun anything.

Our ability to sneak was about to be tested for the first time, and the exam would be graded pass/fail. It is one thing to run your own tests, in conditions where you know every variable even if you pretend you don't. It is another thing entirely to poke around in somebody else's sky, sticking your nose into places someone else is trained to protect. The variables there are incalculable. The guy staring at the radar screen might be asleep, or might be just dumb enough to hit the alarm at something that to a jaded veteran is "obviously a flock of birds." In a practical sense, a bored, twitchy kid can prove just as deadly as a steely-eyed pro, for altogether different reasons.

We knew one of the primary radars guiding the air-defense systems was the Russian-made P-18 radar system. We knew that system well. Designated by NATO as the Spoon Rest D, the radar used a one-meter wavelength that had proved to be a stealth-killer in Kosovo, credited with contributing to the loss of Darrell Zelko's F-117 Nighthawk. From the pushpin on the map between the city and the airfield, the Spoon Rest could watch a 360-degree field of view some 250 kilometers across, up to an altitude of 35 kilometers. We were talking a hell of lot of airspace inside that dome.

If the volume of sky the radar could see was imagined to be a lake, our plan was to walk quietly along the shoreline and see if any alligators jumped out. If we made it around the lake alive, we'd try sticking our toes in the water.

The minutes dragged by as our plane made its way around the outskirts of Kandahar. Although nobody was onboard the aircraft, the GCS had a palpable life-or-death level of stress. It would only take a chat from Roger or other analysts or a blip on our threat warning screen to let us know that our bird was being "painted" by a focused enemy radar, the immediate precursor to some sort of hostile response.

That's unless, of course, Roger had to step away from his computer, an event he would precede with a chat line that read, "No data will be lost." This was in part true because, at least from a purely intel perspective, the entire mission was recorded and could be replayed endlessly, scoured for missed details. But to a flight crew with no rewind button, our survival could well hang on a timely warning. Roger being away also insured that "no data will be gained."

Incredibly, the warning flag never went up. We finished our perimeter sweep, double-checking every sensor to confirm that we remained undetected.

Big looked at me with a wordless shrug. My eyes moved across every screen and across the faces around the room.

"Ginger?" I said on the intercom.

She didn't need me to pose the question, nor did she need a moment to gather data on current threats. She replied instantly, "I got nothing."

"Eric?" Like Ginger, Eric was waiting for the question. "NSA has no unusual comms traffic. It's all quiet."

I looked at Scott, now in the pilot seat, and said, "You are cleared to penetrate the SA-3 ring." Banking the aircraft into the danger zone, Scott took up a course straight through the threat ring covering Kandahar.

An Air Force staff sergeant, Andy R. was a superb sensor operator. He cycled the sensor between the missile site and the alert area for the Kandahar MiG-21s, but we were still too far out to see definable activity at either location. As we got closer, the powerful ball cameras sucked in thermal and ambient energy, constructing detailed images that sprawled the spectrums from starlight to infrared.

A dawning realization spread through the team: *We're in*. We could look down from high ground and—at least thus far—nothing suggested that the enemy had any idea we were here.

I stared at the colorless terrain that scrolled silently across the screen, noting the flow of cars down city streets and the cluster of people who funneled into a building. We had answered the technical challenge and delivered an invisible array of sensors to a point on the far side of the world.

But that was only the first half of the equation. A capacity is only as good as what you can do with it, and the Agency side of the house now had to step up to the plate. We were about to challenge the accuracy of their source data, their human intelligence and satellite analysis, in a way that had never been done before. And we had a man to find.

6: IT'S ALWAYS WHAT YOU LEAST EXPECT

ALEC BIERBAUER

BEWARE OF LAWNMOWERS
August–September 2000

Despite our initial victories, we had to deal with a relentless string of challenges to integrate a new system into the middle of high-priority operations that crossed numerous borders and international authorities. Those obstacles ran the gamut from technical and political down to simple bureaucratic paralysis. Most people would be shocked to find out how easily the president of the United States, while saying "get it done at all costs" could be thwarted by a mid-level manager who hasn't received the proper form in triplicate.

Our transmission task looked like some global game of HORSE: bounce a signal off a satellite of infinitely questionable heritage, through a friendly nation, and down the drain through a trans-Atlantic undersea cable. Nothing but net.

Nobody had even imagined the security protocols needed to get something like that approved. But this requirement paled compared to the challenge we'd overcome establishing a remote-controlled top-secret airfield in the mountains of Asia.

With that airfield up and running, I moved my operations to Ramstein in the summer of 2000. At the risk of suggesting a pattern, our failure-is-not-an-option focus quickly led to some more creative problem-solving. Although we had our Big-Ass Dish, we still needed commercial satellite time.

The subject of fierce competition in normal times, our overlap with the 2000 Olympic Games meant that every second of Ku-band satellite space had been purchased, some segments months if not years in advance. Between live coverage of the events and lucrative commercial slots, the major networks weren't about to give up a prime block willingly, nor had I any intention to explain why we needed it more than the Olympics. Taking a cue from my Brothers in Blue, the straight line to success was to steal it.

Steal is an ugly word and probably a bit of a stretch, although the networks would certainly make the case. It struck me that "unfairly acquired" had a better ring to it. At the time, market price for the block of bandwidth we required was on the order of a hundred thousand dollars. Given how close we were to launch, our mission had reached the celebrated point where "money is no object." The sales rep for the satellite company seemed quite sincere in his regrets that every scrap of Ku-band was under contract, right up to the point that we plunked a half-million dollars on the table. According to our well-placed interlocutors at the Pentagon, he blinked several times, then remarkably found a chunk of available time that had erroneously been listed as sold. There is a lesson in that: technology and tradecraft can solve a great many problems; for everything else, bring a big box of cash.

Now having everything we thought we would need, we set out to run an aggressive operations schedule. If weather permitted, we planned to be in the air three to four times a week.

A lot of that airtime was spent learning how a small plane survives in a big damn sky. Beyond enemy planes and missiles, the very atmosphere itself could prove hostile. Modern radar provides an exceptional look at weather patterns, allowing commercial pilots to go around storms that might endanger an aircraft.

That option went out the window the night we found ourselves staring at a hundred-mile wall of storm that loomed between us and our target. To go around that barrier would have meant trespassing into somebody's sovereign airspace, which carried considerable risk. An American U-2 pilot named Francis Gary Powers had demonstrated the extent of that risk to the whole world by getting shot out of the sky over the Soviet Union back in 1960. That he was flying a CIA spy plane at the time was not a comforting coincidence. Sticking your nose where it isn't invited is the kind of "let's poke the bear" decision that can prompt some very bad, and very unhealthy, responses.

Whereas a crisis to have the smoking wreckage of a covert aircraft splashed across the evening news, that outcome would pale against the diplomatic disaster of having flown said plane through another nation's airspace without permission. In a worst-case scenario, that sort of thing could be construed as an act of war, and no amount of top cover in the world would save us from the shitstorm that would follow. In all of our briefings to the relevant stakeholders, we had never asked for permission to violate the sovereign airspace of any nation—other than Afghanistan.

One option was to call it a night, to turn Predator around and head home. But with the nine-month clock ticking down to the proverbial fourth quarter, the time for do-overs was evaporating. Each mission was predicated on hard-won human intelligence, and the mission we scrub might be the one headed for bin Laden. Due to the weather, we couldn't return on our normal flight route. Plowing straight into the storm and heading into Afghanistan was suicide; all that remained was fudging the borders and praying we were as invisible as we hoped. High in the clouds, Predator slid through the lightning-carved night.

Mark's assessment was concise. After checking with his own threat analysts, he said "nobody else is flying in this shit." From the pilot seat, Big picked his way around the weather and saved it for another day.

Amid all of the effort to make sure that our hosts didn't see our plane coming and going, security back at the MCE remained a comparable concern. A proactive order came down from Mark's boss to put a bit of counterintelligence discipline into play. That entailed mapping out the footprint we occupied and the signature of our activities as others would see them.

It took little more than a glance to recognize that we were an oddity. Unlike a normal DOD team, we were a motley crew of fully uniformed military personnel and civilians moving with equal ease in and out of a highly secured perimeter. We had some major satellite infrastructure; we had pilots but no planes. Anybody associated with our host location would scratch their heads and assume we'd lost our damn minds.

To preempt any uncomfortable discussion about what the hell we were actually doing on their airbase, the Air Force gave OSI[52] the task of running CI[53] against us, to see what we looked like to the outside world. This would tell us if we needed to adjust our profile.

Classic surveillance entails following people, on foot and in cars, as well as the use of various technical tools. As much as we

---

52 The Air Force Office of Special Investigations. It carries an important mission to identify, exploit, and neutralize criminal, terrorist, and intelligence threats.

53 Counterintelligence. CI refers to information gathered and activities conducted to thwart enemy espionage, surveillance, or hostile actions directed at you. It is the essence of cat-and-mouse, trying to hide your actions while watching those who are watching you.

love expressions of loyalty between nations, the conduct of black programs will invariably boil down to "in God we trust; everyone else we monitor." It is less a statement of trust, or the lack thereof, than it is a simple reality in the world of covert operations.

As a part of the CI protocol, we required our guys to make sure they weren't being followed when away from our compound. It was really inconvenient, but we didn't need to associate our people with our location and have our thin cover story unravel. Instead of taking the direct path from one point to the next we are obliged by operational security to adopt circuitous routes, doubling back or pausing along the way. The more we arbitrarily change our daily routine, it is argued, the harder it is for somebody to spot a pattern of behavior.

For some this practice was a matter of discipline, while for others it was simply an unintentional gift. Scott Swanson once demonstrated the latter, the term *Swanson CI* becoming synonymous for driving around town, sincerely and hopelessly lost. On the bright side, his meandering had to drive the local authorities batty if they were watching us and gave OSI the impression that we were astutely practicing top-notch CI techniques.

As the nine-month clock ticked down relentlessly, Murphy's Law continued to challenge our ability to adapt on the fly. Just to keep things interesting, some challenges came in pairs.

One such pair started with the need to troubleshoot the aircraft's SATCOM equipment when we had no L-3 Communications[54] SATCOM folks forward. Since we couldn't just put technicians on a plane to go overseas and fix problems, our solution in one particular case was to ship an oscilloscope diagnostic tool forward to our clandestine airfield, then point the predator sensor-ball at

---

54 L-3 was a major defense contracting company with some unique capabilities including key links to our satellite communications.

the scope so our technicians back in Virginia could see the squiggly lines which meant something to them but meant nothing to our small discrete crew at the forward location.

That wasn't an elegant solution, I realized, but Rube Goldberg would have been proud. At about the same time, the forward team inadvertently destroyed a crypto key needed to run the aircraft GPS system. With a mission hanging in the balance and no means to transmit the codes or ship a laptop securely, we had to revert to old-school tradecraft.

We put Big on a flight to meet an Agency officer flying in from Asia. The objective was a discreet hand-off in the departure lounge of an unrelated airport somewhere in the middle—a place nobody would be watching. Because neither of the two had ever met each other, Big would be looking for an African-American in an orange sweater who answered to the name "Jeff."

Standing over six feet in height, Big was both surprised and amused by the tiny stature of his counterpart. They exchanged a challenge phrase and completed the swap of information. The two unplanned travelers then reversed their steps. As a pilot, not an operative, Big seemed to take considerable delight in his foray into the world of cloak and dagger.

Try as we might to prepare for everything, some hurdles came from way out in left field. In the midst of one mission, a landscaper carefully mowing the grass around the MCE shredded a fiber-optic line. Paul Welch, flanked by his Airmen, labored all night in the pouring rain to repair the cable. Up to that point, nobody thought to include the danger of late-night lawnmowers in our playbook of lost-link mission protocols, but we had one now.

Beyond the wear and tear on operations, living in firefighter mode exacts a heavy toll on the people living from one problem to the next. The ability of a team to maintain a NASA-worthy camaraderie hinges in part on the character and professionalism of its members and in part on their ability to blow off steam in the down-times.

Pranks were a normal part of life, like the time I found myself yelling louder and louder into my office phone in a struggle to be heard. Movement down the aisle between cubes caught my eye and I saw Captain DJ, one of the Air Force liaison guys, almost in tears laughing. I slowly looked at the phone in my hand and saw the layer of transparent tape wrapped across the microphone. DJ would play that trick on me and others again and again.

While unappreciated at the time, the little spikes of humor served as a reality check we all needed. Past miles of cable and layers of carbon-fiber composite, it was the people who made this a success. When all is said and done, we relied on each other more than any arsenal of gizmos. We had all temporarily abandoned our families, our social lives, and possibly our careers to be a part of something bigger than ourselves. We knew how terribly fragile it was; we were never more than one crash or one diplomatic dispute away from being disavowed, maybe prosecuted. Success hung on how hard we worked and how aggressively we manipulated the system. DJ's fucking with my phone became legendary among our ranks, and there is no telling how many phones were ultimately beaten to death on my desk, on one of our trips to far off lands, or in the GRC as a result.

Given the difficulty of our mission, combined with the Type A personalities from the military and intelligence world pitched into the same sandbox together, de-stressing wasn't always a jovial, collegiate affair. We were big boys and girls and once in a while swapping ideas turned into a swapping of fists—work hard, play hard.

Perhaps because of my Army background, I had a penchant for indirect fire. Cube farms are space-efficient but odd in that you work almost elbow-to-elbow with people you can hear but not see.

If I felt the need to make a point, a wadded-up ball of paper lobbed over a cube wall often served as my messenger.

The weakness in that tactic became painfully apparent one night, much like many others, with the team pounding away well into the late hours. I stood up, "prairie-dogging" in cube-farm parlance, down the long aisle just in time to see an arm-whipping blur from Mark's cube some thirty feet away. A tiny copper streak whizzed across the distance and smacked me right between the eyes. I blinked rapidly as the penny bounced to the floor, leaving a dent in my forehead. Mark couldn't stop laughing despite my scowled bitching, but I had to concede his logic. Leave it to an Air Force guy to counter mortar fire with a precision-guided missile.

Each of these experiences contributed to mission-planning and operations: how we avoided detection, conducted operational missions, and dealt with contingencies while coping with human factors. This "work hard, play hard" atmosphere was essential to an environment where the need to make tough, on-the-spot decisions landed in our lap without warning. Decisions like our storm-driven crossing of a sovereign border can bite you in the ass when you are flying a top-secret mission in an unmanned aircraft that on paper doesn't exist.

The how was important, but we never took our eyes off the why. While all of this was going on, we were hunting for bin Laden. The urgency of that mission made every stumble, every obstacle, a matter of great priority. Every limitation was something we pushed against.

One such limit was the "soda straw" video that Predator could provide. From a high vantage point we could watch a mile-wide swath of terrain with almost no detail, or we could zoom in, "looking through the soda straw" in ISR parlance, to see a tiny spot with incredible detail. But when the camera is squinting to read the fine print, you have precious little context for exactly where you are in a particular environment. This took us back to the world of tradecraft.

Our human source network on the ground was getting reasonably good at telling us where bin Laden had been, but they struggled to tell us in anything close to real time where he was right now or, better yet, where he was expected to be in the next few hours. He had been known to frequent places ranging from Jalalabad to Kandahar and Kajaki. Each location was hundreds of miles from the other, and at a sluggish one-hundred-knots air speed we really couldn't adjust all that quickly.

As the technical aspects of the program were being pulled together, the intelligence and analysis portion was also in high gear. Our goal was to tell the Predator where it needed to be in advance of the roughly half-day of transit time to the dozen or so most likely target sites around Afghanistan. If Predator was to grow into a successful hunter, we had to give it one hell of a head start.

We knew from Dave's "Where's Waldo" reporting, along with work that I had done at the Pentagon as early as 1997, that Afghanistan was a Disneyland of sorts for budding extremists. Training camps dotted the landscape. We had a fairly good inventory of the established camps, but new ones cropped up routinely. Periodically, satellites snapped pictures of the camps, and some patterns of activity could be inferred from these moments.

While the analysts were struggling with the adjustment to having God-like control over imagery collection, the ops folks were similarly challenged. Now that we proved the capability and were getting comfortable slipping through radar coverage, we needed to mature the plan to find our target. This fell into the realm of collections management.

During World War II, radio communication was comparatively rare and simple in nature. Although radio equipment was not all that complex, not everybody had a radio. Fast forward to the year

2000, and a majority of the world's population had a cell phone that supports not only voice but also text and other data. With the rise of this capability, a growing percentage of the population, by choice or default, spent a chunk of their day broadcasting where they went and what they said. To SIGINT specialists who pluck data from the transmissions that circle the globe, that smells like a smorgasbord of potential intelligence.

That's not true for Afghanistan, a nation crushed by an 85 percent illiteracy rate and rampant poverty, glued together by a national infrastructure destroyed by waves of war and neglect. Afghans have demonstrated amazing ingenuity to merely survive in such deprivation, but from a modern SIGINT perspective Afghanistan was a black hole. Our most advanced Star Trek sensors cannot rely on signal intercepts in a population where goat-herders live in mud buildings or caves.

Usama bin Laden understood this dynamic. University educated, he recognized the hazards inherent in convenient communication technology. This knowledge, backed by a healthy paranoia, drove him to communicate via spoken or written word carried by just a couple highly trusted couriers. It makes for glacially slow correspondence by todays split-second standards, but it makes for rock-solid immunity to cyber or signals hacking.

If you strip away the many flavors of electronic surveillance, you quickly fall back to the core of traditional intelligence collection and analysis. Human intelligence takes place when formal or informal operatives on your team (or traitors from the other team) gather information and feed in back through a variety of channels, often at great personal risk.

HUMINT can provide unparalleled insight into a subject's state of mind, loyalties, even dreams. Countering that value, it is often speed-throttled, constrained by the security measures needed to protect the collectors or the limits at which data can be transferred.

As a rule, the more we want a communication to remain unnoticed, the longer it takes. Old-school methods like a dead drop might take place once a week—if we miss the window, we are stuck waiting till next week.

This throttling became uniquely critical to anticipating bin Laden's imminent movements. For us to have the lead time to plan, load, and launch a mission, our network needed to give us bin Laden's location in real time, or as close to that as possible. "Where he was" means little, "where he is" is better, but "where he will be" is the Holy Grail.

The prep for each mission began with Mark, who, swaggering like a guy who admittedly had already achieved success in his part, would look at me with a jaunty "Well, Alec, where would you like to fly tonight?"

We had enough time to either park over a single facility for twelve hours or spread that airtime across a couple of different locations. With priority-interest areas scattered along the eastern side of Afghanistan, we couldn't just spend a few minutes at the Derunta Training Camp and then jump to Kandahar because of the hours of intervening transit time.

With the job came the expectation that we would leverage our tools and brains to discern a pattern. We had route planners working assessments of how long it took to travel on Afghan-quality roads between the various known and suspect bin Laden bed-down locations, which in turn helped to assign probabilities of where he would go if we knew where he was the previous night. Our network was strongest in the southeast, so we focused there.

Whereas most people are time- and place-predictable to move between home and work every weekday, bin Laden had no such cycles. Christians might also be fairly predictable on Sunday mornings when they go to their place of worship. Applying that constraint to bin Laden wasn't a great fit either. He was praying five times a day, usually wherever he could lay a prayer rug. For Muslims the best shot would be at Friday prayers, which were usually the best likelihood that he would be visiting with close confidants and possibly at a training camp mosque. We knew of a few. The leading candidates were at Derunta, Khowst, Tarnak Farm, and a place rumored to be in the vast Helmand desert that we had not yet located.

We were sifting through lukewarm leads when we received a report that read "bin Laden was at Tarnak Farm last night." I mulled the one-line report. Tarnak Farm was a walled compound of over sixty individual structures located about twelve miles from the city of Kandahar, less than three miles from the Kandahar airfield. First identified as a bin Laden–associated facility in 1999, it was routinely monitored by satellite with still images as frequently as we could get them. We would see individuals at times but nothing that allowed us to monitor security profiles or patterns of activities.

Tarnak was deemed important to bin Laden because it was at the heart of Taliban-controlled territory, fairly close to the Pakistan border, and close to the suspected location of Taliban leader Mullah Omar. Tarnak was the sort of place where bin Laden might feel safe enough to linger, put his feet up, say a prayer, and maybe enjoy a meal. It was a long shot, but a guy can hope.

THE MAN IN WHITE
September 28, 2000

Preparation for our eighth mission to clandestinely sneak into Afghanistan included a review of our best available intelligence for where we might find Bin Laden. Together with a hard shake of the Magic 8-Ball, we launched the mission which had us winging our way towards Kandahar. On the verge of entering the last thirty days of our nine-month window, success hung on the slim hope of catching up with Usama bin Laden.

Hal and I were managing the mission, and we settled into a lazy circuit in the sky high above Tarnak as the afternoon call to prayer began. More vehicles were in the compound than normal, situated in what looked like a security posture. People had been coming and going, some standing idle while others went about their respective tasks. We couldn't hear the call to prayer, a mournful sound typically piped from loudspeakers perched on poles and rooftops. But we knew when it started.

We had our eyes on a building known to be one of bin Laden's many homes when a man emerged. Looking through the soda straw, we could see that he was tall, dressed head to toe in white. He was greeted with great deference by a small group between the home and the adjacent mosque.

I felt the buzz of adrenaline surge through me like voltage. We didn't need the imagery experts to measure the 6'2" figure to know what we had found. On my screen was Usama bin-freaking-Laden, not a photo from last week or a report from yesterday. From half a world away, we had eyes on the worst terrorist in the world, right here, right now. The most improbable, unlikely, unbelievable words flashed through my mind: *Son of a bitch, we did it.*

As directed by the national security advisor on behalf of the president of the United States, our team had actionable intelligence in less than nine months from receipt of tasking, in advance of the deadline—mission accomplished.

I picked up the secure line and started lighting fires up the food chain. We formally requested a strike package of submarine-launched cruise missiles. Springing up from beneath the surface of the Indian Ocean, the TLAMS should be here in under two hours. We had front-row seats, waiting for Uncle Sam to bring the hammer down on a deadly enemy.

Charlie Allen walked out of his planned meeting after the urgent call from Hal and came down the hall from his office. As soon as he saw the screen, he knew it was bin Laden and proclaimed it emphatically before getting on the secure phones himself to work the next steps to take action. We had notified our chain of command and prepared to support confirmation calls before strikes were launched—or so I thought.

Setting aside all the Jedi mysticism that Hollywood wants to wrap around intelligence training, we really do learn a lot about reading people—body language, how human eyes move when someone is using the left side of the brain versus the right. We're not mind-readers or walking lie detectors, but any intelligence officer worth his salt can tell when something is going wrong. It's tucked away in the little stammers, the funky excuses, the moments of uncertainty as somebody ducks and weaves when he or she should be swinging.

Fielding multiple calls and visits with my chain of command, my CIA "spidey sense" was activated. I had just delivered the report of the century—the miracle Hail Mary pass had just been caught in the end zone with two seconds left on the clock. All we had to do now was kick the extra point to turn "mission success" into "crushing victory."

But the kicker was nowhere in sight. The special-teams squad wasn't running onto the field; we weren't shifting into a kick formation. The coaches on the sideline looked content to run out the clock.

To say I felt sick was an understatement; the knot twisting my guts went way beyond nausea. My eyes were fixed on the screen, on the tall man moving slowly through the group of figures that bowed in his direction.

I walked past DJ, and said, "Get Mark on the handset." I didn't want to say what I was thinking; shit, I didn't want to think it in the first place. I sure as hell didn't want to broadcast it through the GCS.

There was a click, and Mark's voice came on the line, his tone one of anticipation. "Birds away?"

I struggled to say it clearly. "Mark, I don't think they're gonna take the shot."

"Not gonna . . ." half a heartbeat passed as Mark processed the most unlikely words he would hear this day or any other day. "What the fuck are you talking about?" His volume doubled as the impact hit home. If anyone in the GCS had an inkling that things were amiss, the worst fears were now confirmed.

"It's not official, Mark," I wanted to be clear if for some reason I was wrong, but with each passing moment I was growing more convinced. "Nobody's talking with the subs, nobody is double-checking with the shoot authorities. I swear to God, I don't think anybody was prepared for us to really do it."

"Well somebody better get fucking prepared and right fucking now—" Mark tore into an anger-fueled rant that to this day defies print. But his point, no matter how incendiary the delivery, was valid. We'd been tasked with building a new capability and inventing a way to fly it around the world. Done. We'd been told we had to do it in nine months and along the way lay eyes on the most dangerous man on the planet, hiding in a sandbox the size of Texas. Done.

Hal caught my eye, and I put my hand over the phone. "What's up?"

"We are being asked if we can guarantee that bin Laden will still be there in two hours."

My inside voice channeled Mark: *Two hours? We spec'd a Wescam Versatron ball, not a fucking crystal ball.*

But my outside voice remained quiet. "Can we guarantee" is not a question for somebody looking to act; it is a question from somebody looking for an excuse not to act. I could already hear the briefing in my head: *No sir, we chose not to shoot because we couldn't guarantee he would still be there when the missiles arrived.*

Hal came back, anger rising. "Alec, they aren't going to pull the trigger. Can we kamikaze the plane?"

I gave it a thought, a serious one. But as attractive as the idea seemed there was no way to make it work. We'd programmed the Predator to save itself from a crash if at all possible. I fell back on channeling Mark as the word slipped from my lips: "Mo-ther-fucker."

DJ stepped into my line of vision as I refocused on my reply. "No, no sir. I cannot guarantee where he will be—"

DJ waved me off, the urgency in his eyes forcing me to silence.

He plucked the phone from my hand, pressing the mouthpiece against his side as he pointed toward my computer. "Roger's on the chat line," he said tersely. "We have a problem."

7: EVERYTHING COMES WITH A PRICE

MARK COOTER

YOU HAVE THIRTY MINUTES TO LIVE
September 28, 2000

"**T**hey're talking about you."

The words on the chat line had come from Roger at NSA. A brilliant linguist, Roger was a one-of-a-kind resource who could pluck a string of Pashto, Dari, or Urdu out of a garbled radio transmission. At the moment, he was glued to radio traffic buzzing around the Kandahar airbase. His next transmission added, "Yep, they have you."

The intercept from Kandahar threw Roger into unfamiliar territory. His experience focused on extracting intelligence from subtleties of nuance and inflection, not tactical air and air defense. This course of events had just thrown him into a crash course on aerial combat and evasion. I was praying the "crash" part was only figurative. We had experts in air and air defense on our team at Ramstein. This was going to take a team effort. Colonel Boyle and the Lieutenant Colonel "Dash" Jamieson, the 32nd AIS commander, gave us their best analysts. That group included a Mr. Brian "Fish"

Fishpaugh, USAFE's senior air analyst, along with Captain Shane H.,[55] as well as Master Sergeant Rich Keady—all the best analysts.

"They are trying to launch something." A tangible note of tension crept into his words.

"I want eyes on the airfield." I barked, knowing that an order to swing the camera off Tarnak and bin Laden would send the Agency guys off the rails, but I had to know what was about to join us in the sky. The Taliban had an early warning radar with ground-control intercept capability adjacent to the airfield. Just to the west sat another SA-3 surface-to-air missile battery.

A product of Soviet engineering, the SA-3 Goa packed a pair of solid-fuel rockets that would accelerate the missile at or just above Mach 3. We're talking a missile that proved it could run down an F-16. With some 130 pounds of high explosive and frag packed in its nose, the SA-3 needed little more than thirty seconds once it left the rails to close to within a couple car lengths of us. If that happened, our day of success would end as litter spread across the desert floor.

As I expected, Hal was in the GRC and said no on giving up the camera, but I had the benefit of being the mission commander in the GCS. I punched the speakerphone button to get Captain Ty Peterson[56] on the line and said, "Fuck this."

I snapped my attention to Gunny. "Eyes on the airfield, now."

Gunny didn't bat an eye. I could just imagine the cussing all the way back to Langley when our video feed smeared into a horizontal blur as the camera slewed about 140 degrees to the north,

---

55 Captain Shane H. was a towering analyst who specialized in air defense. A USAF Weapons School graduate, Shane was smart, assertive, and a razor-sharp professional.

56 Another one of our indispensable liaison officers, Ty was an Air Force captain with whom I shared prior experience on Predator back at Indian Springs. Ty was wicked smart and cool under pressure, even with Hal breathing down his neck. That alone put him on a very short list.

squaring up on the airfield. In the world of infrared, small grey specks swirled around a dart-shaped silhouette.

"Fishbed," I muttered, the profile unmistakable. While nowhere near as fast as the SA-3, the Soviet-era MiG-21 Fishbed's twin turbojet engines could drive the aircraft in short sprints up to Mach 2. The Fishbed was a veteran dogfighter, proven capable of killing enemy jets with a combination of 23 mm cannon fire and two racks of AA-2 Atoll missiles, the Soviet answer to the American Sidewinder.

I stepped back, my mind processing the options. We had discussed the scenario, planned for it, even trained for it. But nobody in the history of flight had ever done it for real, with the weight of national security on our shoulders. The only thing that could possibly make the moment worse would be to have my boss magically appear, so naturally Colonel Boyle walked in the back door. Our eyes met across the GCS and I'm quite certain he read "aww, fuck" in mine.

God love Colonel Boyle. He just nodded, gave some sage advice, and let me run with it.

Turning back to the screen, I watched a single gray dot scuttle up the ladder and disappear into the jet fighter. The other ants scattered as the heart of the gray silhouette blossomed coal black as the engines roared to life. The Fishbed rolled from the tarmac and aligned itself on the runway. It turned out that Predator looked less like a flock of geese on radar than it did the broad side of a school bus.

Cruiser, another pilot from General Atomics, was at the stick. He saw me get target lock and said, "What's the call, Cooter?"

I looked him in the eye and said, "You have thirty minutes to live. Time to do some of that pilot shit."

The command may have sounded flip, but the reality on the screen was clear, and I was glad to have Cruiser in the pilot seat. If anyone had a shot at exploiting the outermost fringes of our flight capabilities to play cat-and-mouse with a Soviet-built jet, it was Cruiser.

The jet fighter shot down the runway and blistered into the night sky, its pilot's every moment of training focused on hunting other jet-powered objects moving hot, high, and fast. Taking a page from ancient samurai sword master Myamoto Musashi, the perfect response was to be cold, low, and slow.

The tactic was based on a lot more than sixteenth century "meet fire with water" philosophy. The Fishbed had high rails, a design that put the pilot deep in a steel bathtub. He could see well enough ahead and above, but visibility below was for shit. The disparity of speed was a huge advantage to the jet in a fight, but it could be all but crippling in a search. With his foot to the floor in an adrenaline-fueled urgency to find us, the Fishbed could blow by in the night and never glimpse a small, slow speck crawling along the corner of its blind spot.

Gunny once again brought his own special A game to the fight, using the sensor ball in ways for which it had never been designed. The Versatron could spot a dog in the desert or tell if a car's engine was running. The 1,200-degree exhaust from a jet fighter running full afterburner makes it by far the brightest star in the thermal sky. Back in Indian Springs, Gunny had utilized that to follow the Thunderbirds as they trained there.

There was little to see of a jet as it approached, but it shot by us with a Roman candle up its ass. Gunny expertly slewed the thermal camera like a bloodhound on a scent. The smear of heat banked to starboard and swept off to the south. I watched it race away for several seconds before I remembered to breathe.

If we were lucky, we might catch a second break. Most of the Afghans flew under old Soviet protocols in which a highly experienced senior pilot on the ground micromanaged the mission via radio communication with the pilot. Ground control would say turn left, the pilot would execute and then describe where he was and what he saw. With that exchange going on, Roger was gold.

"They're climbing to twenty-five thousand," Roger reported, "swinging to heading two-two-niner."

I glanced at the map. The Fishbed was roaring up into the clouds to the northwest of the airfield. Our bird was below it, headed for the mountains that sloped up to the north.

When I say we had maps of the battle space, it goes far beyond the simplicity of terrain and elevation. Our systems contained a wealth of NSA data on the location, power, and range of every radar in Afghanistan. Our 32nd AIS (Air Intelligence Squadron)/USAFE[57] intel analysts crunched the numbers and calculated every line of sight and every obstruction to produce a three-dimensional virtual space. Our exit point was a dead spot in radar coverage, where the eyes of the enemy couldn't peer over a ridgeline.

If we make that, I thought anxiously, *we become a very small fish out in the open ocean.*

"He's about to come around." Roger declared. "Holding altitude, coming to zero-niner-zero."

"Fuck." The word came out under my breath. A dead-east heading would put the MiG on an intercept vector. But would he see us if he didn't know what he was looking at?

"Stay below and point into him," I said, Cruiser nodding in confirmation. I and the other Weapons School grads on the team were using our "patches" for all they were worth.[58] If the Afghan kept his foot on the floor, his jet would rocket across our path at a right angle . . . about two minutes ahead of us.

---

57 United States Air Forces in Europe

58 USAF Weapons School graduates are referred to as *patches* in the Air Force due to the distinctive "target" patch they wear on their uniform. Patches are the master tacticians and instructors.

The lead-foot pursuit offered another glimmer of hope. At full throttle, the jet-powered Fishbed was sucking down fuel in big gulps, a rate of consumption that would leave him no more than thirty minutes of airtime. I glanced at the clock—this dance had been going for twelve. Eighteen minutes were left, almost the blink of an eye on a coffee break but an eternity in aerial-combat time.

I looked at Cruiser. "Keep turning into him as long as Roger gives us good situational awareness."

Cruiser gave a chuff and settled into a slow left bank, bringing Predator into a nose-on game-of-chicken orientation. But this was no contest to see who flinched. Seen from the side or dead above, Predator is a sprawling sailplane with an outline reminiscent of the old U-2 that Gary Powers flew into Soviet airspace back in the 1960s. That little factoid brought no comfort since Powers got shot down.

But nose-on, Predator almost disappears. At that angle, the view would be like looking at a pair of knife-blades sticking out from a pencil that's pointed at one's face. If our bird was tough to spot from the side, setting it on a collision attitude made it all but invisible—as long as that "collision" part didn't actually happen.

The undertones in the room blurred as the seconds ticked by. Roger was pumping a nonstop stream of Afghan ground control into chat. Captain Shane H. quickly analyzed the raw information and passed it to the GCS in an operationally focused manner.

The MiG-21 roared by us, high overhead. Roger relayed that the pilot couldn't see anything.

A minute went by, maybe two, as we watched the MiG disappear in our proverbial rear-view mirror. Nothing else looked to be scrambling; we weren't being painted by a hostile radar. I gave a sigh as I picked up the handset and asked for Alec.

"So what now? Do we head back to Tarnak?"

Alec sounded like a guy who just lost his dog. "No, we're done. Everything upstream is, I dunno, disengaged. Whatever it was they thought they'd have to deal with tonight, it looks like *success* didn't make the list."

"Rog that." I set down the phone.

Turning to the flight crew I gave Gunny a tired pat on the shoulder and said to Cruiser, "Well, that was fun. Head for the hills."

SO WHAT DO WE DO NOW, BUTCH?
October 2000

Over the next few weeks, Alec and I remained in the grip of numbness and frustration. We continued to fly our docket of missions, running through the paces of collecting intelligence, documenting patterns of life.

I found myself wondering why we bothered. Patterns of life help you get breaks on key intel. Key intel puts you on a high value target. But we were on the Ace of Spades of all HVT, and for all intents and purposes we gave him a little wave to say, "Gotcha." We had no justice to share with the families of the embassy-bombing victims, nothing encouraging to tell US citizens—hell, no warning to the next miscreant who wants to blow up a building full of Americans. I found myself living in Despondence, the capital city of Miserable Bastardland.

But self-pity wasn't my style, and military training gave me nothing if not the ability to self-generate initiative. Whatever indecision that contributed to Usama bin Laden still drawing breath occurred way above my pay grade, above the team of patriots that had sequenced miracles to put that son of a bitch in the crosshairs. Nobody understood the why of it all; we all just knew we still had a job to do.

Alec and Hal were working in the shadows to determine where to place blame for the Mission 8 no-go and shore up support for making sure it never happened again. Alec had the relationships with the J3 for the Joint Staff from his Pentagon days. That led to some squirrely, off-the-record comms to heavy hitters at the NSC to build an amended plan. Because Alec didn't wear a uniform, it was easier for him to jump a dozen layers of rank to the top of the food chain. The fact that CTC had issued him a shitload of rope to hang himself with didn't hurt either, as long as he kept his head out of the noose. Alec continued to try to convince me, and maybe himself, that we could still make shit happen.

Like each step forward, our efforts revealed yet another unexpected challenge: how to ingest so much data and cherry-pick the immediate highlights while sorting and storing the bulk for endless laps of subsequent analysis. Even though the task was daunting, it was a good problem to have. With the Mission 8 letdown, we now apparently had time to work some of these issues. From the inception of the program, the management of the data was overseen by Steve, a full-motion-video (FMV) guru from NIMA. That might seem misleading, however, because there was next to no FMV work taking place there, or in the CIA.

Regardless, Steve from NIMA was writing checks that were making policies and procedures on the fly. In just eight flights we had amassed a data set that had become hard to manage, much less exploit for intelligence production.

Alec, the team, and I were intent on fixing things to prevent another failure to engage. We had a video conference with our team to get everyone back on the same page. Alec wanted to take everyone's pulse and assess their resolve to keep going. Alec explained the events of the 28th, that we had successfully put US technical eyes on target and did so inside our nine-month window. He explained that as a part of the Pentagon's commitment and at the NSC's urging, America had parked submarines in the Indian Ocean, quietly turning circles off the coast of Pakistan. They had been waiting, somewhat patiently, for us to give them something to shoot.

Submarines were being used instead of surface ships because they could get closer without detection. Closeness means shorter the flight time to target. Tarnak Farm, along with a dozen other facilities, had been preprogrammed into their targeting computers because we felt Tarnak was a high-probability target location. As a bonus, Tarnak was also the southernmost target and closest to the submarines. Plus, we spotted the man in white during Friday afternoon prayers in Afghanistan. Bin Laden would almost certainly have a meal and sleep there after prayers.

Having all nine planets aligned, somebody up the food chain was afraid to give the order—afraid perhaps that bin Laden would suddenly up and leave, afraid of . . . well, it didn't much matter. It boiled down to the old axiom: no guts, no glory—and no elimination of Usama bin Laden.

During our meeting we brainstormed how we could further shorten the delivery time for a missile strike or remove impediments from the firing sequence. But while the physics of speed and distance were immutable, it had become clear that government will was unreliable.

Having overcome a hundred prior obstacles to get to this point, we fell back on what had become our unofficial motto: "Never mind, we'll do it ourselves."

POPULARITY IS A FICKLE BITCH
October 2000

In the wake of September 28, the "whatever it takes" commitments began to fade. This became clear when people suddenly started haggling over the bill. What had been "all in" suddenly became "all yours."

The FCF crash proved to be one of our most prickly issues. The Air Force adamantly maintained that it crashed in the hands of Agency and General Atomics, which put it squarely in the realm of "you break it, you buy it." The Agency swung back, claiming that the Air Force was at fault for not accounting for the operational constraints of night flight. At the rate things were going, cooperating engineers would be replaced by conflicting attorneys.

At that point, we had still not written down who owned the damned aircraft in the first place. The entire program was grinding to a halt as we struggled to figure out who would get stuck with the $3 million tab.

Though we were technically still operational, we routinely had missions cancelled at the last minute—some as we were on the end of the runway ready to takeoff. Though little more than transparent brinksmanship, it was nonetheless very frustrating.

Our forward team was suffering from a lack of information and purpose. As we fought through the bureaucracy inside the beltway, they sat idle on the far side of the planet with minimal support and a strong desire to be the pointy end of our program. This team of General Atomics contractors and Agency personnel represented the smallest footprint we could design. An Agency officer in charge, medic, and communications officer were supporting four contractors: a pilot, two mechanics, and a technician. After that, it was a strong liaison relationship with our partners for security and logistics support.

When our DC based budget or policy challenges hit, one of the first phone calls was to the officer in charge forward to stand down, perform maintenance, and standby. While not proud of how we treated the team, it was a necessity. The people would still get paid and were safe, but, like us, they had come too far in this program to sit idle.

Perhaps because of the out-of-sight, out-of-mind feeling or simply being so far removed from the decorum of the beltway, the LRE team found increasingly creative ways to make its insolvency a point. As video feeds from each of the increasingly rare Predator takeoffs streamed to Langley and the Pentagon, the camera would sweep over messages spelled out on the runway in thermally active lettering. Phrases like SEND MONEY followed by JUST DO IT became the Predator-equivalent of Burma-Shave signs. Our higher-ups did not appreciate the humor. All the while, bin Laden would continue to plan and execute terrorist attacks.

Alec briefed me that after much debate, CIA Executive Director Buzzy Krongard called all the interested parties into his office and declared that all the attempts to professionally resolve this stalemate had been ineffective. Absent some basis for accounting, he fell back on the tried and true: we should go "halfsies." Remarkably, that settled it. The Air Force and the CIA each kicked in a million-five, and the team got back to work.

After September 28 we flew six more missions. In at least five of our fifteen missions, we ran up against MiGs, with two close encounters. But that tiger had lost its teeth. The lessons learned over Kandahar quickly led to further improvement in tactics.

One of those missions took place around Kabul when we got word from Roger that a MiG was being launched. We were too far from the airfield to see the jet aircraft so we did the only thing we could do: run. Big turned us north into the highest peaks of the Hindu Kush. It was one of the few times I've looked up at terrain in a Predator. The turbulence coming from the mountains started increasing to the point we were at risk of losing our satellite link. After a chat with Big, I instructed him to climb. I feared the turbulence more than the MiG and figured if we were going to "die" I wanted to see it happen, not disappear in the frozen video screen of lost link. Once again, we made it home.

SHIT, WE'RE A TOM CLANCY NOVEL
October 2000

One of our last missions of 2000 focused on scouting Gharmabak Ghar, a relatively new addition to the database of training camps in 2000. Skeptics in the intelligence community offered up dismissal that it was an NGO facility or some other benign entity. We believed it to be a training camp of some sort but couldn't determine critical aspects like structure, discipline, or rigor. Without clear evidence, everything was mere speculation. Alec needed to confirm what his clandestine sources were reporting, and the satellite snapshots weren't getting us there.

That's not a knock on the NIMA guys. Although that bit of space-age technology gives America incredible advantages, it carries some inescapable vulnerabilities. A satellite is not a plane that can fly around at will—it's a rock orbiting the earth at a fixed speed, largely on a fixed track. That means it will pass over a given spot at a given time—no sooner, no later.

Sadly, at this point the orbit of these satellites was largely known by sophisticated adversaries. The timetables for some satellites were even posted on the internet. As a result, people on the ground could adjust their outside activities to avoid detection when satellites were due to pass overhead. The Russians and Chinese were masters at that game, having elaborate denial and deception practices to avoid being seen or to create misleading visible signatures.[59]

---

59 To be fair, we were doing the same thing. The US has a rich history of "tactical deception" that includes the celebrated US Army 23rd Headquarters Special Troops. An elite team of artists and illusionists in the midst of a shooting war, the "Ghost Army" staged over twenty fake battlefield scenes during WWII, suckering Luftwaffe pilots to report sighting entire platoons of inflatable tanks and scarecrow soldiers. Enemy assets dispatched to chase ghosts were not shooting at our very real flesh-and-blood soldiers.

We didn't necessarily believe that al-Qaeda was quite that competent, but it made sense that the group was clever enough to run inside when the satellite was due to pass.

By this point, the GCS team was practically running itself. Relief teams showed up ahead of shift and the transitions went like clockwork. We were familiar with our routes, with our procedures. As mission commander, I found myself leaning back and watching excellence at work.

Basking in the luxury of spare bandwidth, I wondered if things had gelled as well back at home. Alec had a much larger group, divided among more diverse pedigrees. The GRC was a mix of communications technicians, mission managers, imagery analysts, collections managers, SIGINT analysts, and liaison officers from multiple agencies and organizations with some sort of equities in the program. Had they grown together, or had the mix created frictions? As we settled into our orbit over Gharmabak Ghar, I pulled on a headset, cranked up the volume, and punched the line to Alec.

"We're on station," I announced, knowing that Alec was already glued to the video display. What we saw was breathtaking.

Most of the locations we had watched to this point had been largely benign. The airfields had some points of interest, but by and large, most of the residential locations were just that, mud homes where the only movement outside might have been a goat wandering in the yard. By comparison, Gharmabak Ghar was a terrorist Disneyland.

It took just moments to determine, beyond any doubt, that we were not watching relief workers but troops in military formations. Trainees were called out to do push-ups, formations were sent out on a group run. We could count heat signatures inside of tents.

Three individuals caught our eye, running side-by-side for short lengths then stopping in unison before advancing again. Then another group followed suit. On a hunch I said, "SO, go back to IR."

The feed switched to shades of gray, and we could see the unmistakable jet-black thermal flicker of muzzle flash from the business end of AK-47s—advancing fire drills.

"I don't think those are NGOs, Alec, but hey, what do I know?"

Across the open line I heard Alec's attempt at a touchy-feely approach. "OK, boys and girls, we worked hard to get here, and you now have control of the sensor. Tell me what you want to see that years of satellite imagery has failed to provide for you."

I was shocked by the instant onslaught of "look here, no . . . pan left" and "show me that from the other side." By the sound of things, one would think somebody just threw a bucket of fish guts in a tank full of sharks.

"Cripes, Alec," I said, unsure if he would hear me over the din. "Are the kids excited today or what?"

"Today?" he came back, his tone a bit incredulous. "Dude, this is every mission. To these guys you're the freakin' crack dealer."

I blinked, my eyes looking as if for the first time at a screen that had become all-too normal for me, something that perhaps I had taken for granted.

The NIMA team was a meticulous group that had been sorely neglected by the intelligence community. These folks until now had survived on meager rations: a stack of photographs each day, most from satellites, backed up by a handful now and then from some recon aircraft skipping along the upper edges of flyable atmosphere. These very hungry minds suddenly had a full-motion Thanksgiving dinner dumped in their laps—so much food that nobody knew what to bite first. If something looked promising, they didn't have to wait a day hoping for a better satellite photo. They simply asked us to steer the plane and zoom the camera to squint at any given target with laser focus. The chat line scrolled like a stock-exchange ticker.

Of course, every new capability brings new challenges, and I realized I had no idea what sort of effect our tsunami of visual data would bring. Storing gigabytes of data, tagging it for search—the list was endless. But based on the cheers and high fives I could hear in the background, it was a problem that the team was anxious to tackle.

"They feel like they've been chucked into that Clancy movie," Alec said, his voice a mix of pride and amusement.

I knew the reference—*Patriot Games*. In that story, Jack Ryan was a CIA intelligence officer charged with tracking down a terrorist cell in north Africa. When he did, British Special Air Service (SAS) sent a direct-action team to assault the camp as Ryan watched via satellite feed to CIA headquarters in Virginia. In the scene, the monitors showed live "thermal" video, fancifully portrayed in shades of blue. The frame rate and clarity were so vivid you could see bodies change color as the slain terrorists cooled after death.

A big hit when it came out, most of us thought *Patriot Games* pegged the bullshit meter when it came to surveillance technology. At the time, we were living in a world where experts struggled for hours, sometimes days, to declare some shapeless blob on a still image to be a person or weapon. Live thermal satellite video wasn't just a stretch, it was utter fantasy.

I slowly shook my head. That was 1992. What a difference a few years can make.

While fiction for the time, the movie visualized a phenomenal capability. The ability to watch a target before, during, and after an event would be a huge leap, peeling back the fog of war that has harried military leaders throughout history. While Hollywood was certainly not a motivation for Predator, it rather accurately predicted the clarity of intelligence that would be required to get decision makers to approve action in Afghanistan.

Listening to the headset I found myself wondering if Alec felt like Jack Ryan staring down at Gharmabak Ghar. Then a sense of pride kicked in for what we had accomplished. We were cheaper than a satellite and could circle a target for hours, and our video wasn't restricted to shades of blue. Frankly, we were better on every metric. I couldn't suppress a tiny grin of pride. Suck it, Ryan.

A TRAGIC HOMECOMING
October 14, 2000

Scott and I were traveling west on Kisling Memorial Drive, headed for the left turn that would take us to our site on Ramstein. Spending so much time virtually in the desolate browns and grays of Central Asia, the beauty of Germany seemed almost alien. Looking at the rolling hills of green, we could almost forget why we were here.

I glanced to my left where the two huge C-17 cargo planes sat alone on the wide Air Mobility Command ramp. This was not unusual; it was a very busy part of the airfield. However, the anomaly was a row of gloss-black limousines that stretched from the tail of the aircraft. No, I caught myself; it was a row of hearses.

As the column pulled away I could see several vans at the tail end, speaking to the number of the dead—more bodies than hearses to carry them. I felt my gut sink, realizing instantly that it had to be the seventeen American sailors who lost their lives in the bombing of the USS *Cole*.

Scott slowed to a stop in the roadway, coming to rest just short of the turn. "Let's get out," I said. He followed me, and the two of us stood in silent salute as the somber procession left the base, turning east, doubtlessly toward Landstuhl Regional Medical Center. We held our place until the last vehicle drove out of sight.

We climbed back into the car, the muddy swirl of anguish in my gut quickly settling into a tightly coiled ball of pissed-off. We screeched into the parking lot, blew by the guards, and made a beeline for the secure phone.

I barely heard Alec say "Hello" before I cut loose with both barrels.

"What the fuck are we doing?" The conversation, a monologue really, went downhill from there. I ranted, raged, vented about having to stand in a roadway as our dead kids drove by. Thankfully, Alec just took the brunt of my frustration as it came full circle, calmer at the end, spent.

"What the fuck are we doing, Alec?" I repeated where I'd begun, this time a sincere question. "What's the point of finding a guy like bin Laden if we can't kill him and shit like this happens?"

The answer wasn't what I wanted to hear. That's not to suggest that decisions for an Air Force officer are black and white, but they rarely run a hundred subtle shades of gray. We have rules of engagement, and we play nice right up to the point that somebody crosses a clear line; then we blow them out of the sky or bomb them off the face of the earth. The Air Force mission is not long on subtlety.

The CIA, intertwined across the worlds of intelligence, defense, and politics, is another beast entirely. It lives in the gray zone. Alec reminded me that it wasn't our fault, that somebody in Washington wasn't up to making the call. The decision wasn't on us.

Talking me down out of the proverbial clock tower, Alec reminded me of the discussion that had played out in the GRC when we were sitting 16,000 feet above bin Laden. At the time, we had grappled for anything we could do to bring lethal force into play, going so far as to debate if we could just crash the whole damn plane into bin Laden. Sadly, that option was thwarted by the very safety measures we had built into the aircraft—it would probably go lost link before it hit the ground.

I had heard it all before, but in the wake of watching those flag-draped coffins roll by one after another I just couldn't stomach the idea that we had found America's greatest threat but, with the world's greatest arsenal of firepower on our side, we couldn't find anybody with the balls to shoot the son of a bitch. I wondered if we could have prevented the *Cole* bombing by killing bin Laden that night at Tarnak Farm.

At some point my ability to talk just ran out of gas. The phone felt heavy in my hand. I was sick and tired, and I knew Alec felt the same.

"So what are our options?" Alec posed the question seriously, his tone helping me to steer away from the frustration.

My brow furrowed, and thinking was difficult. "We could leave it where it is," I said, "fly it as long as we can. But winter is rolling in and conditions will go all to hell. We'll be lucky to get one day in ten up in the air."

"I concur," Alec said, adding his own affirmation. "The LRE was never meant to be a permanent base. We don't have the long-term authorities in place." He paused. "And more of the same would only produce more of the same results."

The idea of that chafed my nerves, and the growl edged back into my tone when I said, "Shit, Alec, I'm about ready to strap a bomb or missile under the damn thing and do it ourselves."

I grimaced as quickly as the words came out of my mouth. The desire to punch back appealed to the American in me, but the Air Force guy knew that strapping any kind of weapon onto a Predator would raise a hundred flags, technical as well as political. The very idea bordered on madness.

8: A DEADLY IDEA

ALEC BIERBAUER

TABLE FOR FOUR
October 31, 2000

Frankenstein's monster leaned a green-skinned elbow on the bar, pitching his best line to a blonde in some revealing nurse outfit. Another couple walked past in head-to-toe black and sunglasses, presumably recent escapees from *The Matrix*. They say you can run into just about anything in Washington, DC, and certainly never more so than in DC's upscale Georgetown on Halloween night. After much deliberation on a costume, I showed up dressed as a CIA case officer—nondescript attire that to the untrained eye might have looked like the same clothes I'd worn to work.

Star Trek's Captain Jean-Luc Picard emerged from the crowd in a gold Federation shirt and smiled when he saw me. The man behind the Starfleet pin was Dick Clarke, whose October 31 birthday was the raison d'être for this gathering. He walked over and shook my hand, falling easily into casual social banter about the characters that filled the room. His tone shifted when we were joined by Roger Cressey and Mike Sheehan.

Dick had invited me to the gathering as a likely thanks for sneaking Roger and him into CIA headquarters in the wee hours of the morning to watch missions over Afghanistan. They had the proper clearances, but normal protocols and notifications to the front office should have taken place if everyone was playing well in the sandbox. Regardless of the reason, I was pleased to have the face time with some like-minded people who sat in very influential positions.

Cressey had a pedigree that spanned the Departments of Defense and State, with a wealth of expertise on Middle East security. He had served overseas with the United Nations peacekeeping missions in Somalia and what had been Yugoslavia and had time at the US Embassy in Tel Aviv. Sheehan was the State Department coordinator for counterterrorism with the rank and status of ambassador-at-large. These guys understood all too well that the real monsters don't wear capes and plastic fangs.

We found a quiet corner at the back of the bar. Everybody was aware of the Predator program, and all supported it. Commiserating in our drinks over the stall in activity, we shared the question, "What more could we do?"

It was just one week before an election that the polls declared a toss-up between the two most polar-opposite candidates, George W. Bush and Al Gore. Despite the immediate uncertainty and the huge political letdown, we maintained a low-level resolve to keep the program alive—even if that meant pushing it forward as a grassroots effort.

As the discussion meandered, I waited until the notion of arming Predator surfaced on its own. The concept had popped up once or twice before, the question largely dismissed as academic due to the numerous restrictions and ample number of challenges just running a split ops mission. But we had put many of those challenges to bed in the past nine months. With so much of the impossible behind us, the question of weaponizing Predator floated from "Could we?" to "How?"

The Air Force had a science project on the shelf to kick off some five to six years down the road, to experiment with arming a UAV.

It was little more than a couple lines in an out-year budget projection, but it was enough of a spark to light our fire. We quickly identified the key people needed to assess weaponizing our surveillance system. A quick discussion with General Atomics and the Air Force revealed that the wings could only hold about one hundred pounds of weapons each. Normally delivering firepower by the ton, the Air Force didn't have anything that small in its inventory. Our next call would be to the Army to see what it had.

The notion of an armed UAV had been floated before, at least at the concept level. The idea harkened back to the early days of aerial combat when World War I pilots took to the sky in planes made of wood and cloth. Long before anybody had the idea of bolting on a machine gun, pilots figured out that gliding above the battlefield gave them a unique angle from which to shoot. Some began to carry pistols, engaging in little more than pot-shot wars with similarly inclined opponents. Others, however, carried hand grenades and ultimately small bombs that were chucked out over enemy lines. Those were small measures to be sure, but in the right place even a tiny explosion can have a big impact.

Regardless of our renewed enthusiasm to move forward, we knew the odds remained, as always, stacked heavily against us. We were now taking an already fragile platform even further from its intended purpose. The coming weeks would be crawling with lawyers and accountants, leaving precious little time to meet with the engineers, mechanics, fabricators, and computer guys needed to do the actual work. For every step forward, at least one new impediment was in our way. We were desperate but determined.

The tall pole in the tent would be getting a mandate from the new administration, whatever administration that happened to be. Short of that massive turnover, there were short-term options like getting top cover from DOD and intelligence-service stakeholders to advance the underlying science.

And there was a hell of a lot of science that stood between our cocktail napkin and a lethal aircraft. Despite the progress we had made, we had no idea what tool of destruction we could duct-tape to a remote-control plane. Another flurry of engineering would be required to figure that out, even more so to make it all work.

The four of us bandied the options as an exercise in open-ended brainstorming. At a more gut level than any sense of engineering, I was leaning toward a laser-guided bomb as opposed to a missile. My logic was simple: with about a hundred pounds of carry weight on each wing, or a single 250 pounds strapped to the belly, every ounce of motor and fuel in a missile was weight that came off the high explosive payload. If a 20-pound missile warhead performed well at the receiving end, a 250-pound bomb would be better. Since gravity was a free resource and there was no rush in terms of the "drop time," my thought was to strap the biggest ball of explosive we could carry under the belly and let physics handle the rest. Very few kinetic options cannot be enhanced by adding more high-explosive material.

Setting that aside, however, choosing the weapon itself was only part of the equation. Whatever munition was selected, it would require some sort of pylon or rack to carry the weight of the weapon, the release mechanism, and all of the electronics to make it run.

The wings would need to be redesigned as well, as nobody had envisioned them with weapon hard points. There'd be software to write, integrating a whole new line of operation into our existing command-and-control infrastructure. If the political fallout of crashing a remote-control airplane seemed career-ending, one could only imagine the impact of that flying toy erroneously dropping a bomb on some mosque, school, or supermarket. The weight behind the guidance "don't fuck it up" grew exponentially.

No hard requirements resulted from the Halloween Table of Four, but we each left with a renewed sense of purpose. Dick had to sell it downtown, and at the time, he was unsure which political party would be in office and what if any role he would retain in the new administration. Sheehan had some work to do at State, and I had to sell to a human-intelligence-gathering organization that our next step was to launch missiles. I didn't have a full appreciation of the technical aspects involved but would soon learn just how complex creating a new war-fighting capability would be.

A major political challenge was that the outgoing Clinton administration had failed to seize the opportunity on Mission 8. The Pentagon, while ready with missiles in the launch basket, failed to convince the administration that the relative probabilities were as good as they were ever going to be. Finally, the CIA failed to take the capability seriously before Mission 8. The administration had issued the challenge, the Pentagon had the technology, and the CIA had the authorities. Despite long odds, we had marched the ball deep into the red zone and, with less than a minute to play in the game, we failed to punch it in. Regardless of all our accomplishments, it seemed everyone involved had come up short at some level. Common sense suggested that nobody would welcome the idea of being put back on that hot seat.

While the political side of the effort was large and amorphous, the proverbial exercise of wrangling cats, the technical side was also sprawling across a wider footprint of entities and authorities. That led us back to our call to the US Army.

A hundred-pound weapon seems far more substantial when you have to heft it up under an aircraft wing. But on the scale of weapons the Air Force normally delivers, a hundred pounds is a rounding error. A single JDAM, a mainstay of aerial bombardment, can run from as light as five hundred pounds to a hefty two thousand, and even at that no aircraft carries just one. An F/A-18 can carry some thirteen thousand pounds of ordnance, and an F-15E can haul a staggering eleven and a half tons at supersonic speeds. Going back to Vietnam, the venerable B-52 tore through the skies with some thirty-five tons of bombs onboard, hitting speeds about six times that of Predator.

The Army, however, is a service that revolves around the lone soldier, and what that soldier can carry. Every additional bullet or band-aid comes at a cost that often must be offset by a little less of something else. As a result, the Army had made a science of thinking lighter and leaner. Weapon systems like Javelin and TOW had given the foot soldier a tremendous ability to project precision force from a shoulder- or tripod-mounted launcher, but those weapons carried barely a soda-can's worth of explosives. A five-pound warhead can punch a lethal pencil-hole in tank armor, but it is not designed for wide-area effect.

Everyone involved took on the challenge to find the right capabilities for this mission. Air Force, Redstone Arsenal, and the Army were all involved, along with the CIA's internal group within SAD.

As we rifled through stacks of manuals and spec sheets, it became clear that we needed to bring in a ringer. Mike P. was a veteran of the CIA's Special Activities Division, which did the heavy lifting for most of the Agency's covert and paramilitary operations. Mike had an extensive knowledge of Army systems coupled with a well-earned reputation down at the AMRDEC[60] at Redstone Arsenal in Huntsville, Alabama. Mike quickly

---

60 The Army's Aviation and Missile Research and Development Engineering Center. A primarily civilian organization, AMRDEC is filled with actual rocket scientists tasked to provide research, development, and engineering technology and services to support US Army aviation and missile platforms.

identified another tank killer that looked like a solid match, the AGM-114 Hellfire.

Emerging from the dawn of helicopter warfare, Hellfire was designed to provide greater punch and far greater accuracy than the point-and-shoot folding-fin rockets that lit up the jungles of Vietnam. With an overall weapon weight of just a hair under one hundred pounds, the Hellfire packed a full twenty-pound warhead.

Twenty pounds still struck me as anemic. I clung to my internal leaning for ninety pounds of C-4 and ball bearings stuffed inside a ten-pound can. But Mike and the engineers in Huntsville were adamant that the Hellfire was a good fit. It was designed as an air-to-ground weapon, even if originally intended for much lower launch altitudes. Hellfires were readily available in the current inventory and had existing launch rails, and we could grab mechanics familiar with the system.

The per-shot cost of about $70,000 to $100,000 was trivial on the scale of things, a nice benefit for the taxpayers, especially when compared with cruise-missile costs. As a bonus, the Hellfire was a laser-guided weapon, able to home in on a dazzling point of coherent light the likes of which we had in the new MTS ball from Raytheon. Hellfire had already racked up an impressive service record that stretched around the world from Panama to Yugoslavia.

Pictures of Hellfire, along with schematics and tech specs, joined the growing montage that sprawled across the wall like centerfolds from *Popular Mechanics*. Assuming that one decision held, the choice of weapon could set a great many additional dominos into motion. The weapon gave us a direction for the weapon rack, in this case a single-rail Hellfire pylon designed for an Apache gunship helicopter. Knowing the rack gave us the hard point requirements that we could backtrack to Big Safari and General Atomics to integrate into a new wing design.

The ad hoc evolution staggered forward in surges, marked in places by sidesteps and the occasional faceplant. But if a constant thread ran through the program, it was one of misbegotten heritage. Reminiscent of the multinational satellite months ago, we were looking at taking a surveillance drone and hanging Army missiles on helicopter racks bolted to one-off wings. Once again, our bird was becoming more of a mutt every day.

9: THE BREAKING POINT

MARK COOTER

RESIGNATION
November 2000

We had planned to cease operations due to the weather, but higher-ups kept having us fly when we could. One of those just finished, another long one. Normally, I would arrive about four hours prior to takeoff and stay until we landed. That can make for a twenty-two-plus hour day with a combat nap here or there.

We had accomplished what we had set out to do, to provide actionable intelligence on bin Laden. In addition, we had expanded our knowledge on al-Qaeda operations in Afghanistan and, for that matter, the Taliban forces. It had become obvious to me that, since we were past the election, no action would be taken regardless of the intelligence we provided. I was exhausted; what little sleep I got was haunted by the image of hearses driving by—a line of hearses that never stops. The weather was closing in, flight ops were done, and I was done.

I sat down at the computer and through bloodshot eyes typed a letter to Alec, Colonel Boyle, and Snake. The gist of the letter was we had done everything we were asked to do, the weather wasn't getting better, and we were at higher risk of losing an aircraft and thus compromising the program without having taken action against bin Laden. Continued operations seemed senseless. We should pack up, go home, and regroup until we had the assets and resolve to take action. I ended with something to the effect "if the decision is made to continue operations, I will resign my commission in the United States Air Force." I hit Save and named the file "My Resignation Letter."

I asked Ginger, someone I dearly trust, to read the letter. Her eyes got wide, and she said, "Holy shit, I think you should have Dash read this before you send it." Dash was the 32nd AIS commander, who I'd known since 1987 when we worked on the A-10 together and trusted implicitly.

Ginger had never steered me wrong, so I called Dash and asked her to review the letter as well. She was fiery, smart, and aggressive. She immediately came over and sat down in front of the screen. For the first time in my career, I was way more aggressive than Dash.

Her reaction was very calm, yet firm. "Okay, so let's hit Save. Ginger, get someone to drive him to his room." She then looked me in the eye with a stare that brooked no debate. "Get some sleep, and let's review this again tomorrow."

I'd seen that look before and knew when a fight was lost before it began. And Dash was right, I'd need a good night's sleep if I was going to chuck my career.

The next day, we got together and reviewed the letter again. We toned it down just a bit, deleted the part about resigning my commission, and punched Send. Thank God for Dash.

I think that Alec's leadership got nervous about the prospect of DOD bringing my reservations up to the White House. So the Agency preempted DOD, and I was told the letter made it to the White House in four hours. I wasn't sure if that was true, but two days later I was on a plane to DC to debrief the operation and discuss the next phase.

10: BUILDING A KILLER

ALEC BIERBAUER

REDSTONE
November–December 2000

The chamber of commerce in Alabama claims that Huntsville has the highest density of PhDs in the country. I don't know about the accuracy of that statistic, but walking around the prestigious Redstone Arsenal, I just about couldn't wave my arms without hitting a rocket scientist.

Redstone is home to a number of tenants, including the US Army's Aviation and Missile Command, DIA's Missile and Space Intelligence Center, the Missile Defense Agency, and NASA's Marshall Space Flight Center. One would likely be hard-pressed to find anything rocket-driven that couldn't trace some part of its heritage through Redstone. A full-size Saturn V rocket that towers over Space Camp[61] along Interstate 565 near the Arsenal's northern gate suggests that this is a source of some pride to the residents of Huntsville.

We called on those experts for some hands-on with Hellfire, meeting with some of the big brains in Army missile systems. One

---

61 Space Camp is an educational camp on the grounds of the U.S. Space & Rocket Center museum at NASA's Marshall Space Flight Center.

of those, Michael Schexnayder, was the associate director for systems at the Army's Aviation and Missile Research and Development Engineering Center, the Army's focal point for aviation and missile platforms. Together with the Joint Attack Munitions Systems Program Office, these two entities represented perhaps the best path to a viable weapon system.

Michael was incredibly bright and politically astute, with his finger on the pulse of the missile community. It was immediately clear that he, and the team he represented, had probably forgotten more about rocket science than we would ever know.

As a side benefit, Michael had an almost photographic knowledge of the workings of Redstone: who held what position and what political lines could impact support. He knew everybody firsthand and often knew the predecessors to the current job holders. Needing to get the Army to play nice with the rest of the kids in the sandbox, Michael's institutional knowledge proved as valuable as the technical.

Michael brought in Chuck Vessels, "Boom Boom" to his friends, a strapping 'Bama boy with a disarming southern drawl. A second-generation engineer, Chuck focused less on delivering a payload than he did on maximizing the havoc wrought when it arrived. In a world of shaped charges and frag distribution, Chuck reaffirmed my sense that if a small explosion worked, a bigger one would excel. I'd found a kindred spirit. Chuck became the face of the dozens of patriots at Redstone working to make Hellfire a viable tool for Predator.

To counter my naive skepticism of the missile being the right answer, Mike sent Didi, "The Black Widow," to meet with us. With a handle that begged images of a Soviet assassin, we were surprised to meet a gracious, slight woman who, on the verge of being a grandmother, just happened to spend her day job modeling weapons effects against equipment and people. Didi specialized in determining how many lethal fragments would come off a given weapon, and in what directions, given a specified angle of impact and a variety of materials being impacted.

It was with a chilling degree of relish that she explained the virtues of Hellfire with the help of a visual aid called "bug splat" charts. At first glance they seemed aptly named and likely created by taping paper to the grill of a truck and driving the highway on a warm summer evening when the mosquitos are thick. But here, splats marring the radial pattern of lines and rings indicated the hypersonic passage of white-hot metal. Some frags would punch through steel or a given thickness of concrete. Even the weakest ones would shred the hell out of flesh and bone. I made a note to myself that if I ever saw Didi start to draw chalk lines on the ground where I was standing, I would run for it.

She walked us through the science within a blast, which to that point I had only thought of in terms like "shit blows up." It turned out to be far more complex. While aspects of explosive carnage can be measured in units of temperature, overpressure, or circular error probability, the most important one to a missile program was simply PK, an unassuming little acronym for probability of kill. This is the scientific assessment of what happens when Didi's bug splats happen in the real world. It begins by assuming delivery of the weapon accurately to the circular error probability of the specified weapon—in short, that we hit the target.

From that point, PK is a numeric representation of the statistical probability that the weapon will detonate close enough to the target with enough effectiveness to destroy it. The value is a number from zero to one, so a small fraction like 0.1 infers rather light damage and decent odds of surviving. But 0.8 or 0.9 indicates significant damage to the target. The odds of walking away from a "point-nine" are pretty dismal.

The Holy Grail was a 1. That meant a 100 percent fatality, nothing survives. I might have been mistaken, but I'd swear I saw a gleam in Didi's eyes when she described a 1.

What became clear was that Didi was a singular genius who saw order and distribution in the chaos of an explosion. She and the team in Huntsville did an amazing job, and I came away a believer in precision missiles.

Our sudden abundance of knowledge about Hellfire triggered yet another round of choices, the first being one of models. The product of extensive refinement at Redstone, Hellfire came in a series of flavors tailored to meet the Army's evolving requirements. The K-model carried a shaped-charge warhead designed to punch through a Soviet tank. A newer M-model existed that was equipped with a blast/fragmentation warhead designed to use against urban structures. It was unclear which would perform better against a "thin skin" like a car or pickup truck, or what Didi referred to as "troops in the open." To our mind, it was conceivable that a missile designed to hammer through an armor-plated tank might zip thru a Toyota like tissue paper and not notice.

Altitude was another freshly opened can of worms. Designed to be launched from a helicopter, Hellfire had never been shot from the kinds of altitude we needed. If the temperature is sixty degrees at sea level, the ambient temperature at twenty thousand feet is a frosty ten below zero before factoring in things like moisture and wind chill. Air pressure drops by half at that height, adding to conditions that the missile had never been tested against. Nobody knew if it would come off the rails like a comet . . . or fall off like a frozen rock.

Lending no comfort at all was the subsequent call from Terry McLean at Redstone. He relayed the official word from Lockheed, manufacturer of the Hellfire, that hauling the missile up to twenty thousand feet would void its warranty, so let the buyer beware.

Lockheed got credit for building the base missile, but that response let us know we were unlikely to get support from the factory.

That last point was the least of our worries. The Army had sufficient quantities of missiles on the shelf, so we really didn't need Lockheed's production line. Boom Boom and the team at Redstone by comparison had earned a seat at our table as comrades in arms, joining our growing band of hand-picked renegades and patriots to set aside organizational and institutional stovepipes in the interest of doing bad things to bad people.

With all that in hand, I had to sell Redstone to the Agency, necessitating me to pitch as close to a sure thing as possible. My bosses were no more rocket scientists than I, and they sure as hell wouldn't sit through a lesson on bug splats.

I asked Cofer what he was looking for in terms of probability of success. As you might imagine, he led off with "100 percent" but later moderated that back just a smidge. Smug with Didi's sage teachings, I jotted "PK .95" in my notebook.

Back in my office, the smug fell away as I looked at that brief notation like a college kid staring at his calculus final after an all-night kegger—not that I'd actually known what that felt like. Whatever the case, it was time to put on my math hat.

According to Didi, the number had to pass through several gates. First was the effect of the missile coming off the rail, combined with the probability of the missile seeing the laser, further combined with the probability of the missile working properly from an altitude at least twice as high as it was ever intended to be used. That number bounced off the odds of the missile actually hitting the right aim point, before finally running through Didi's bug-splat calcs on the odds of fragments hitting targets in the right place. Two pencils and three erasers later, that exhausting equation boiled down to a percentage somewhere in the low- to mid-seventies.

To an organization traditionally charged with recruiting spies and stealing secrets, I had to sell what in the best of cases looked like a C- grade, which most of my ivy-league, bowtie-wearing colleagues had never seen in their lives.

On the plus side, I had the Air Force in my corner. It had confidence in a legacy system like Hellfire, certainly over the alternative of trying to build something from scratch at the eleventh hour. According to Redstone, creating a new weapon system normally took years, if not decades. We didn't have that kind of time. Somewhere was another USS *Cole*, another US embassy, and another somebody planning to blow either up.

I looked down at the math, the weight of that pressure on my mind. My eyes parked on a final probability of success that fell somewhere between 70 and 73 percent.

I circled it, paused, then rounded up to 80. Calling it a 90 percent probability seemed too much of a stretch. Collecting my thoughts, I set off to find Rich Blee.

As usual, Rich responded with bottom-line practicality. "I don't give a damn about the math, Alec. Just tell me, will it work or not?"

I exhaled hard, thankful for the chance to make my case honestly. The math might be accurate, at least over a thousand shots, but the odds of getting hit by any one of those points of failure right out of the gate were low. The odds against multiple simultaneous failures was even more remote. Ultimately, Hellfire was a known weapon system boasting a terrific track record. With a clear conscience, I looked him in the eye and said, "Yeah, it'll work."

Rich shrugged, accepting my assurance on face value. "Then let's go sell Cofer."

The meeting with Cofer went well, although it took an altogether unexpected focus. Having predicated his expectations at 95 percent, Cofer seemed far less concerned about the chances of success as he did the impact of failure. The question that boiled up was, "Can we survive bin Laden surviving?"

It was an interesting question, one that I admit had not crossed my mind. If we took a shot at him and missed, he would certainly know he'd been shot at. We weren't talking some errant sniper's bullet that might pass by unnoticed. Bin Laden might be a sociopath, but he was no idiot; it wouldn't take much thinking to put two and two together as to who put him in the crosshairs.

My first-blush thought was *so what?* Bin Laden was already doing everything one human could do to wage war against us, blowing up ships and buildings. Were we really worried about pissing him off? But we went through the possibilities. Maybe he does a Gadaffi, Libya's terrorist president who got permanently rattled after Ronald Reagan bombed him in his own home. In that case, a near-miss was almost as good as a kill. Maybe we would injure bin Laden, sending him into a spiral of eroding health that leaves him incompetent or dead down the road. Maybe he would go to ground, which would at least hamper his ability to command and control al-Qaeda in a timely manner. The worst-case scenario was an emboldened bin Laden using our failure to strengthen his position and standing with his followers.

We sifted through a dozen outcomes, but none of us could project a reality worse than a healthy Usama bin Laden moving freely about Asia.

Cofer nodded and said, "Then make it happen, and don't miss."

I left the meeting with a surge of empowerment and an unfamiliar sensation that I dimly recognized as hope. While our expanding team drove forward, doing our best MacGyver impersonation, I wondered if the rest of the political gears were turning behind the scenes in Washington. No matter what miracles of technology we were able to sequence, none of it would matter without the authority to pull the trigger.

The change of administration came with some relief, not because I was a Republican but because we had seen the risk aversion of the last administration and there was hope that the new administration had a greater tolerance for risk. I hoped Dick Clarke would stay in place along with his team and get the new administration to seriously look at al-Qaeda. I had not counted on the transition events or timelines that started with the need to find the *W* keys that the exiting Clinton staff pried off every keyboard in the White House, the vandalism a sophomoric, if not outright criminal, parting gift for George W. Bush.

On January 25, 2001, Dick Clarke presented a memo to Secretary of State Condoleezza Rice that emphasized an urgent meeting of principals was needed to review the al-Qaeda network. In what would prove to be the mother of all understatements, he pointed out that al-Qaeda was not some "narrow, little terrorist issue" and went on to recommend budgets and authority to develop measures to counter the growing threat of Islamic terror. Dick had done his part and continued pressing forward as one of our biggest advocates. The political pendulum began to swing.

CALLING DOWN HELLFIRE
December 2000–January 2001

Blazing the trail of vetting Hellfire on Predator was Air Force General John Jumper, leveraging the Air Force Big Safari team. The team had fought its way through a brief turf war with a competing program office, so on a formal green light Big Safari director Bill Grimes wasted no time in tasking three Air Force majors, Brian "Radz" Raduenz, Ray Pry, and Mark "Spoon" M., with introducing plane to missile out on the test range. It was a great choice; more qualified guys were few and even farther between.

Their task had not been an easy one. A hundred questions stood between the first scribble of Hellfire and Predator in the same sentence and actually making the idea work. High on that list of questions was a basic go/no-go sort of query: Will Hellfire's launch rip the wing off a Predator?

Answering that question required a partially disassembled Predator looking like a forlorn beast chained down to a concrete pad. Its severed wings sat off to either side, perched on aluminum scaffolds. A launcher was bolted under one wing, or part of one at least. What had started life as an M299 launcher now had three of the four rails cut away. Hanging on that lone rail was a Hellfire.

This first missile was just one step above a dummy round, having neither warhead nor main motor. The neutered weapon—a "house mouse" in Big Safari parlance—needed only the launch charge to punch it off the rail. Unseen was a handful of sensors that would measure the forces exerted when the missile fired. If the test went well, those faint but meaningful measurements would be critical in determining how close it came to the edge of the envelope. If the test went poorly, well, the threshold would be obvious.

Three miles downrange sat the remains of a tank that had seen better days. Although massive up close, the roughly sixty-ton vehicle looked like a speck at that distance. Like magic in the late morning sun, the missile disappeared in a sudden flash. It streaked past the sound barrier, hurtling across the divide like a cat sprinting for a laser dot. That dash ended when the missile slammed into the turret, splattering like a metal egg.

Big Safari's test proved our most dire fears to be unfounded. The missile didn't rip the wing off, didn't set it on fire; it didn't seem to even scratch the paint. The test was then repeated, this time with the wings actually on the plane, producing the same result.

The final step had been to fire a live missile, complete with a working warhead. Like the shots that preceded it, the missile scorched off the rails, sailed over three miles of test range, and slammed into the metal hulk. Instead of a thud like Thor's hammer, the warhead detonated with a flash, adding a burning hole to the two fresh dents.

The next tests would see if Predator could perform an encore while actually flying. The test plan called for the plane to launch from Indian Springs and fly to the test range where the pilot, Air Force Captain Curt "Hawg" Hawes, was waiting along with sensor operator Master Sergeant Leo Glovka. It had been a brief series—three shots, all taken from two thousand feet AGL.[62] As before, all three were fired at a dead tank. The ascending level of difficulty between each shot was spread across a number of technical factors.

The first shot was buddy lased, the missile homing in on a dot of coherent light projected by a separate laser. Control of that shot ran over an LOS link, the line-of-sight connection by far the most reliable. Absent a warhead, the Hellfire splattered itself against the metal hull.

Shot two was still inert but required the missile to self-lase. That meant projecting its own dot and keeping it on target while the missile hurtled in. To double down on the complexity, control of the missile was switched from line of sight to Ku-band,[63] banking the signal off an overhead satellite. Both aspects would be critical to effectively engage targets when the nearest ally was

---

62 Above ground level. AGL indicates a relative altitude above the ground directly beneath, rather than an absolute height above sea level. Each scale has a particular value. Three thousand feet absolute might tell you all you need to know over the ocean, but AGL is all-important when flying toward a mountain.

63 The Ku-band is a portion of the electromagnetic spectrum in the microwave range. Simply put, it is the bandwidth most commonly used to communicate with satellites.

in another country. That missile punched yet another dent in the target, leaving only the addition of high explosives to round out the trifecta. When the final missile blew a hole through the tank's turret, there was little doubt in the base capability. What lingered, however, were questions about how it would all hold together at altitude, when the complexities of wind, cold, and distance expanded geometrically. That would send the test team to China Lake.

IRON MEN AND MUD HUTS
January–February 2001

While the shots against tanks continued in the desert, in the back of my head I was still stressing how well Hellfire would perform in different target scenarios. If we played out our "man in white" sighting from September 2000, an outside shot would likely have been lethal, but what if he had wandered into a building? Could we penetrate the thick mud walls?

In the old saying "close only counts in horseshoes and hand grenades," the former is due to a basic rule of the game, and the latter holds true as a matter of physics. A hand grenade is designed to fill an area with fragmentation—bits of its own case hurled with explosive force. Imagine thick chunks of razorblade moving at bullet speeds. A grenade that explodes in your lap might kill you from blast alone, but the shockwave might not be lethal just a dozen paces away. In contrast, a face full of shrapnel at that distance can end your day.

That basic concept applies to any sort of detonation and harkened back to my persistent desire for more explosive. The layman might think that a warhead capable of killing everybody inside a tank would wreak havoc on a mud hut, but as I had learned at Redstone, the net effects were likely to be very different.

Antitank warheads use shaped-charge technology, a bit of physics known as the Munroe effect. Instead of exploding outward like one might normally imagine, a shaped charge is designed to focus the power of the blast into what one might visualize as a lightsaber of energy that burns straight into the target, a force that can lance a neat hole through steel armor.

When that happens, all of the molten metal—along with additional bits that spall off the inside surface of the hull—are launched with incredible violence. Unable to punch through the opposite side of the tank, these bits carom all around the cabin, like a swarm of white hot, supersonic bees that are really pissed off. These bees will punch fiery holes in equipment and people, as well as very touchy things like fuel tanks and ammunition magazines. Secondary explosions and gasoline-fed fires can rapidly transform the crew cabin of a tank into a blast furnace.

We had no firsthand knowledge of what the same missile would do when it hit a mud ceiling. Concrete shatters but mud . . . squishes? We were fairly confident that the lightsaber part would punch through a mud-and-beam roof as easily as through steel, but we feared that most of the force would just pass through the airspace below and drive itself into the ground. Those standing just few feet away from the point of impact might get a hell of a scare but little in the way of grievous injury.

Over the years, the Army had run penetration tests against just about every sort of building material: earth on timber, block, brick over block, poured concrete, all of them studied, measured, and indexed so niche experts like the Black Widow could readily model what a given weapon will do against a given fortification. The challenges of our air war in Iraq focused on penetrating eighty feet of steel-reinforced concrete in buried underground bunkers.

Historically, people who lived in mud huts hadn't posed much of a threat to the western world. That was conventional wisdom, which prevailed before suicide bombers drove a ramshackle fishing boat full of explosives against the hull of our billion-dollar guided-missile destroyer. As the threat of Islamic fanaticism crept closer to American soil every day, mud huts became relevant.

My bosses had the same concern. How would we know the effects in the all-important last twenty feet of the thousands of miles kill chain if we didn't try it? Whereas the need to answer the question was great, the how looked a little sketchy. Looking across this great nation of ours, we couldn't come up with anyone who had real-world experience building mud huts the likes of which cover Afghanistan. We couldn't begin to guess what consistency of mud was used, or the ratio of mud to straw, or the role of manure in the recipe. It was time to call in the big guns—Alex Lovett and Chyau Shen.

Regardless of the depth and breadth of the Agency reach, when I absolutely needed something done quickly and without a lot of scrutiny, I had come to depend on my new friends at OSD and NAVAIR 4.5x Special Surveillance Program. Lovett had proven his weight in gold for navigating the Pentagon to get us a direct path to folks like Mark Cooter and Predator. Lovett had also introduced me to Chyau Shen and his unique lab of patriots at Patuxent Naval Air Station. I really didn't appreciate it at the time, but these American heroes had enough resource and expertise to prototype miracles.

Chyau was already working other parts of the program after the initial Charlie Allen meeting, including a ground-based system that emerged from our early desire to have unattended ground sensors monitoring key facilities. While we weren't comfortable getting ground sensors around places like Tarnak Farm, we were excited with Lovett and Chyau's concept for how to monitor the Derunta training camp outside of Jalalabad.

Derunta was one of the most northern facilities where our source reporting indicated bin Laden would visit. Just outside of Jalalabad city it was safely off the main road across the Derunta Dam and in the foothills of the Hindu Kush mountains. The facility was routinely used as a training, test, and evaluation location for explosives and chemical weapons. One of the discussion points at the January 2000 meeting in Charlie Allen's office was how to develop a persistent technical monitoring capability. As we aggressively pursued Predator we knew we couldn't be all places all the time and would need another source of intelligence from sources on the ground, other technical devices on the ground, or phone intercepts to give us an indication of where to we should send Predator.

Lovett and Chyau, together with Doc Cabayan, proposed a long-distance telescope to be parked on the mountain overlooking Derunta as a unique capability to monitor this particular camp. We sketched out the concept, and the experts went to work on a technical solution while Hal and I started working the plan for how to get the tool in the right place.

The mountains north of Derunta were largely a no-man's-land between the Taliban and our close allies in the Northern Alliance. Our terrain analysis and discussions with Ahmad Shah Massoud and the Northern Alliance revealed an option we could get to from the north that would put us over ten thousand feet elevation about forty kilometers slant range away from Derunta. When we revisited the requirement with the Lovett and Chyau, the math revealed a need for a very specialized lens to determine men with weapons from that distance.

Because of the remote location, the system would need to be self-sustaining for months and would need to be able to remotely pan, tilt, and zoom to monitor the area. Because of the location, it would also need to be carried up the mountain by men or mules. My ideal size and weight package requirement was a naive forty to fifty pounds to be carried by one to two men.

Lovett and Chyau came back with a custom built, tripod-mounted telescope packing a remote-controlled gimbal connected back to CIA headquarters by Inmarsat satellite with both solar and wind-generated power stored in marine batteries. Yes, the actual package needed eight to ten men to carry it, but life was full of trade-offs. It was still a miracle.

But at this point I didn't need technology magic; I needed an Afghan mud hut. Silicon and titanium had to give way to a mixture of mud, manure, and straw.

Chyau's answer was brilliantly simple, and he called in Steve Steptoe. Steptoe was the entrepreneurial understudy to Chyau and one of his go-to guys to get things done under extreme constraints and timelines. Steptoe was also the one to build the remote telescope destined for the mountains of Afghanistan.

Chyau and Steptoe set off south of the border to Mexico and returned with a literal truckload of new construction techniques and materials—the closest representation of construction material and expertise that money could buy. They headed for China Lake to construct an accurate representation of bin Laden's Tarnak Farm house.

As with so many challenges along the way, we identified the problem, brainstormed options, found mom-and-apple-pie-loving Americans, and got out of their way as they created a solution. Chyau and his team were determined to turn off the challenges from Washington regarding weapons effects against a building.

Chyau was never one to skimp on a task, and this was no exception. We didn't have direct access to bin Laden's Tarnak Farm house, but we did have anecdotal discussions from Afghans familiar with the area, as well as the best measuring and assessments our NIMA and DI analysts could produce. We had to guess on a couple items, like the roof materials, but how far off could we be, right?

MEANWHILE, BACK AT THE RANGE
May–June 2001

It's said that those who don't learn from history are doomed to repeat it. The uncomfortable irony of that nugget felt a little too close to home when we drove out on the empty flats of China Lake test range where, not but a year ago, we had shown up with a crazy idea about sending a pilotless plane into the spy business.

If you set aside Roswell, New Mexico, or Nevada's legendary Area 51, few places carried more tinfoil-hat conspiracy theories than the Naval Air Weapons Station China Lake. Rich with its own supposed UFO lore that stretched from lights in the night sky to gargantuan holes in the desert floor, China Lake was a highly classified remote test facility. But unlike Area 51, where experimental aircraft are shoved out of the nest to see if they can fly, China Lake is where the most gifted fledglings find out if they can kill.

Located in California's Mojave Desert, China Lake's 1.1 million acres sprawled across almost twenty thousand square miles of empty, rocky desert. To put that into some sense of scale, China Lake is bigger than the state of Rhode Island. Add a couple thousand buildings and a couple thousand miles of roadway and you have one hell of a test environment. The sky overhead, about twelve percent of California's total airspace, was restricted to outside traffic and tightly controlled.

Though arguably on the low end of the lethality spectrum, we were assigned a remote corner of the range with what seemed like an infinite amount of sky to ourselves. On the face of it, we felt like we had graduated to the credible, to a seat at the grown-up table. In reality, it was just as likely we'd been banished to a place so far from mankind that if we had a runaway missile, it wouldn't have enough fuel or energy to make it back to civilization.

The China Lake test plan called for thirteen additional shots with a mix of inert and live missiles. Building on our success from two thousand feet in phase one, Big Safari picked up in stride by scoring direct hits from five thousand and eight thousand AGL. By that point, it was hard not to swagger. The bird was no F-16, but it proved the ability to fire an antitank missile out of the sky with what looked to be amazing reliability—right up to shot three, when we shanked one off the fairway and out into the proverbial tall grass.

I experienced a moment of stark clarity when my range officer announced that our missile is tracking off target, followed a heartbeat later by an ass-puckering need to know exactly where it was heading. The rationale for our being in the middle of nowhere suddenly shifted from undeserved exile to a pretty damn good idea.

As it turned out, the missile didn't run far astray, slamming into the desert floor. But in missing the target, the missile scored a direct hit on any creeping sense of infallibility we might have formed. It reminded us all that rocket science was, well, rocket science for a reason and that every minute of every day, Mother Nature was conspiring with the hidden flaw to send our science project to its grave.

The hidden flaw in this case lay not in aerodynamic surfaces or lenses or actuators but in miles upon miles of computer code. The culprit was identified by the real rocket scientists back at Redstone as a PRB error, military-speak for pitch relief bias. In fact, most of the Hellfire code resided in the head of Larry Nuzum at Redstone and his colleague Dave, who was recently retired from Lockheed but had been pulled in to support Larry.

When there was a problem or a tweak needed, it wasn't Lockheed on speed dial; it was Larry and his team. They worked magic on the missile's computer code. In describing the PRB, Larry explained that because Hellfire was designed for comparatively low-altitude launch, inherent in its design was the reflex to go nose-up to get some altitude while the engine was in its three-second burn. This also allowed the missile to avoid terrain in the way of the shot, since Apache helicopters normally flew nap-of-the-earth: at very low altitude, hugging the terrain. After the engine burned out, the rest of the flight was inertia versus gravity.

Larry had been involved since the inception of contact with Redstone, and his early recommendation was to angle the launch rails down five degrees to help compensate for the higher altitude of Predator when compared to an Apache gunship.

But as our launch heights increased, the PRB reflex began to fight with the math of the big arc, the trajectory line that curved from the Predator to the target. Once the missile jumped too high off the curve, it lacked the ability to maneuver back. The Hellfire itself had a limited vector within which it could correct. Although the K-models in use at the time were superior to earlier versions, radical steering was not one of its claims to fame.

Larry and his programmers scrambled to change variables and add subroutines, a mad flurry of wand-waving that promised to fix the problem right up to the point that shot four also went astray.

One miss is a problem; two in a row is a pattern. Recognizing the undeniable reality of "three strikes and you're out," the alpha nerds stepped up, driving a relentless cycle of iterative simulations. Revisions to the software came out nightly, sometimes more than once a day. But the problem proved resilient in the simulation runs. I watched, waited, and paced as days ticked by.

The sun rose on May 29. The programmers had toiled for six days and six nights, and while Genesis tells us God created the earth in that time, we all had our doubts about a working piece of software.

The stakes could not have been higher: you can hit your target time after time with little more than an "attaboy," but three straight misses and somebody will run down the halls of the Pentagon declaring you a failure and wanting your funding—and maybe your head.

On the morning of the 29th, some ten thousand feet above the desert floor, a Hellfire punched off the rails. I don't think anybody spoke for the thirty seconds it took the missile to fall out of the sky. The room was silent until the missile slammed into the curved turret of the target tank. While the warhead was inert, the explosion of cheers and high fives filled the gap. The team, led by Larry and Dave, came through with the latest in our sequence of miracles.

We had three more shots to take, all from twelve thousand feet, and one more from fifteen thousand. Of those, only one missed, due to a mundane missile malfunction not considered systemic. The upshot was clear. From just under three miles up we could drop a missile inside a circle the size of a Volkswagen Beetle.

That left just one remaining question: we could hit a target, but could we kill it? We set off to check out our new mud hut quietly baking under the California sun.

WELCOME TO TACO BELL
June 2001

"Hoh-lee shit."

A couple of the guys who climbed out of the Suburban echoed my response; others just whistled. I had asked for Tarnak Farm. But looking at the fortress of mud before me, I felt like I'd been given the Alamo.

That impression grew as I walked inside, noting two-foot-thick walls that could likely hold off a battering ram. Overhead, the ceiling was crisscrossed with twelve-by-twelve beams set on what seemed like twelve-inch centers. The thought of punching through this mud vault seemed a bridge too far; as I looked, I began to wonder if our twenty-pound warhead would even dent it. I could just imagine the tail of a Hellfire sticking out of the roof like a golf flag.

Spoon looked at me and chuckled. "Chyau go a little heavy on the mud?"

I shook my head, unsure what to say. My gut told me it was likely less a matter of Chyau overengineering than it was the NIMA guys having to guesstimate the wall depth from photos of Tarnak Farm. But whatever the cause, it was the only mud hut in town.

Someone along the way came up with a fanciful cover story for the existence of the oddly placed building, going so far as to hang a colorful sign that read HELLFIRE TACOS on the side of the alleged eatery. Almost instantly it collapsed into TACO BELL, and the name stuck.

Plywood silhouettes, "witness panels" in bomb parlance, were scattered around the scene to capture the effects of blast and frag. In a good case, a missile shot would leave them perforated with holes. In a great case we'd find nothing but splinters. The concept was familiar to anybody who has watched an episode of *MythBusters*. The patterns of effect would be carefully measured and mapped in three dimensions. That data would be back into the Black Widow's sophisticated modeling simulations.

Even though all this work was deadly serious, suggesting that a bunch of Type-A personalities loose in the desert with a stack of missiles will produce consistently reverent behavior would be a lie. On *MythBusters*, the two mad scientists routinely added ballistic-gel mannequins or pig carcasses fitted with high tech sensors. We couldn't afford that, so we placed watermelons atop our wooden silhouettes, the shape, size, and gooey red centers an arguably excellent stand-in for a human head. In the spirit of one-upmanship, somebody ran off a stack of life-sized photocopies of a human face and dutifully taped one to each melon. Any resemblance to Snake Clark, or anyone else, was purely coincidental.

We based the distribution of stand-ins, both in and around the building, on the footage from Mission 8. One specific target stood in for the Man in White. In addition to the humans-in-the-open we parked a number of thin-skin vehicles around the building to get an understanding of how well they would survive. All that remained was to shoot the damn thing.

We launched from ten thousand AGL, a height at which we had hit and surpassed over a week ago. With my entire focus on the question of penetration, Murphy's Law reminded me who really runs things in the world of aerospace. The missile veered off into the weeds.

The glare on my face drew a rapid-fire series of shrugs, disclaimers, and a flurry of finger-pointing. Nobody wanted to own the bragging rights to missing the broad side of a barn along with the entire roof. A flurry of forensic examination dropped the hot potato of blame squarely in the lap of the tracker head. That was a manufacturer issue.

I scowled. Lockheed had told us that taking a Hellfire that high would void the warranty. *Let the buyer beware.*

As disappointing a start as it was, we had nothing to correct. The only path ahead was to load up another missile and send Predator back up to ten thousand feet. Another countdown, another interminable delay, and boom. The missile streaked down and slammed into Taco Bell's wide, flat room. A geyser of mud and shattered wood blasted skyward.

But did it get through? With no outward evidence we could see on the cameras, the thought was at the front of everybody's mind as we raced out to put our own eyes on target.

Smoke wheezed out of the open doors, thick with the smell of burned earth. Through the haze inside the building I could see a thick column of sunlight shining down from the ceiling. The interior was splattered with a mulch of earth and watermelon. I walked past plywood silhouettes, some still upright despite holes I could stick my thumb through.

We took one more shot at Taco Bell, punching a hole through an exterior wall as effectively as we had through the roof. The series of tests wrapped up on a smaller static target, adding yet another kill to our tally. Our scoreboard stood silently in the desert, a gutted building surrounded by perforated cars and pulped melons. Part of me would always hope for more, a giant smoking crater or a rolling mushroom cloud, but those effects were never in play to begin with. What was doable, what had started out on the frayed edge of the possible, was now solidly in the proven. Chyau's overengineered structure left us highly pleased and confident we were viable against this target type. Clutching pages of notes and a stack of photos, I headed back to Washington.

SHOOT/DON'T SHOOT
June–July 2001

Beyond the technical challenges facing this program were the ones created by our media-saturated world. It has oft been opined that the West could not have won World War II with today's media, that given the losses suffered on Normandy, pundits would have declared the war unwinnable on the evening news and beat that drum until the will of the nation eroded. In World War II we heard the likes of Tokyo Rose take to the air telling us to lay down our arms and accept defeat. Today we have MSNBC.

While that might sound like a bit of snark, the political realities of 2000–2001 were clear. Outcomes were measured not only against humanitarian principles and Law of Armed Conflict but also against what I termed the Washington Post Test, which asked, "What is the worst possible way this outcome can be construed and run on the front page?"

I was comfortable with physical challenges. Machines and electronics run on logic and physics and for the most part behave predictably. Things get a lot fuzzier when you have to rely on concepts like courage and intelligence in the topsy-turvy universe of politics.

The elephant in that room continued to be our willingness to shoot. The last time I looked at Usama bin Laden we had submarines full of missiles parked for the express purpose of blasting him off the face of the earth—with no decision to do so. I was determined that we would never again fall prey to lack of commitment.

If simulation worked for modeling combat, perhaps it would work for modeling a combat decision. By presenting the decision-makers with a few likely scenarios, we could build some precedent upon which real-life decisions could be based. Lieutenant General John "Soup" Campbell, along with one of the senior mission managers, gave us the vision to turn this into a practical exercise.

My goal for the tabletop exercise was to get some clear marching orders, criteria defined in the calm of day, that set the bar for use of force. Our team drafted up three scenarios in declining levels of clarity. The first was intended to be the no-brainer: almost to the point of "Usama bin Laden is standing alone in the middle of an empty desert; do we shoot or not?" Given that all of this had been put into motion to achieve that end, it seemed to be the easy start.

The next two scenarios increased in complexity, introducing lower levels of confidence coupled with higher risk for things like collateral damage. My presumption was that I'd get a "hell yes," "maybe," and "hell no," in that order. But pushing a policy question up the chain of authority is a lot like pressing the fire button on a missile for the first time: I never really know what is going to happen.

With scenarios in hand, we were able to assemble the deputies or principals for each relevant element that could, or would, have a seat in the decision-making process. This included Joint Staff generals, the CIA counterterrorism chief, the deputy director for operations of the CIA, the National Security Council counterterrorism coordinator, senior Air Force generals from the Joint Staff and Air Staff, the associate director of the CIA for military affairs, a senior Department of State deputy to the secretary, and a representative from the Office of the Secretary of Defense. Most had at least one lawyer in tow.

Opening with an insincere but very respectful representation of the decision-making community, we moved quickly to the no-brainer vignette. One by one the attendees acknowledged that this was a best-case scenario and voted to shoot. Or so I thought until I realized that one very important pair of palms lay flat on the tabletop. The stated reason was that there was not a 100 percent guarantee of a successful kill and the consequences of failure were too unpalatable to assume the risk.

I was taken aback but on reflection not entirely surprised. What we all saw was most likely a microcosm of what played out as we stared at the Man in White. Nonetheless, for the purpose of this exercise, the no-vote was a substantial blow to the gut.

Undeterred, we pressed forward, receiving the expected results for the remaining two scenarios. We left with a clear understanding that I still had some institution-level selling to do. In my heart, I prayed that our Seeker of Guarantees wouldn't be the one making the call when next we had the world's most wanted terrorist in our sights.

11: CONNECTING THE DOTS

MARK COOTER

REMOTE SPLIT OPS
Summer 2001

We were hurtling towards an epic wreck of conflicting actors, authorities, and national sovereignties. It might be lawful to fire a gun in our own yard, but do it from a friendly neighbor's yard, especially without their permission, or hell, even knowledge, and we end up with a shitstorm on our hands. That problem gets a lot dicier when we stop plinking at tin cans and start shooting at real people.

No matter how many ways we ran the tabletop exercise, the unanswerable questions remained. First, do we tell our partner nation what we've really been doing from its backyard for the last year? Nobody saw that as going over well. Second, are we now comfortable upgrading those missions to potentially lethal strikes? The kerfuffle to doubtlessly follow could spill into the media, sending the element of surprise down in flames with a dozen or so careers. That was a shit sandwich, and nobody wanted a bite.

Word came back that the CIA was unwilling to declare the program to anybody. As usual, Alec's timing sucked, and I let him know it.

I had my hands full at the Pentagon when I received a note from Ginger with a progress report from her end. The tents were gone. We now had a beautiful building with real toilets. She was happy.

And I was about to change that. Given the latest bit of analysis, it was dismally clear that the MCE was going to have to move and much of Ginger's hard work was about to be undone. I dreaded the call but needed to tell Ginger to stop work there and start buttoning up anything that could be moved for shipment and a start-over somewhere else. That call wouldn't be pretty.

True to my prediction, she let me have it. When she asked, "Where are we going?" I could only say, "I'll get back to you." I hung up bloodied and bruised.

I scrolled across the world map, thinking out loud as I scratched one nation after another off the list of candidates. We knew where we wouldn't go but not where we would. Absent any clue on that score, if we were going to have a chance at success we needed to keep the momentum and keep everything running in parallel.

"So what are we talking here, Alec?"

Alec and I had our fair share of head-butting, but his tone remained calm as he drew me back to the positive over several minutes of give and take. For just a moment I felt a twinge of thanks for what I perceived as kindness before I remembered whom I was talking to. Alec was doing that Jedi case-officer shit, walking me to a place that I was beginning to think he had already picked out. I'm no CIA guy, but the poker player in me came to the front.

My eyes narrowed. "Who have you been talking to, Alec?"

It was his turn to chuckle. A second or two ticked by as he considered what he wanted to share.

"I had a meeting with Albert." Alec pronounced the name with a hint of reverence. My brow raised with recognition—the legendary Man with Two Brains.

Through the course of developing the Predator Hellfire, we'd had no shortage of extremely impressive personalities and patriots. But without a doubt, Albert was senior among the brilliant.

He'd been brought in as a subject-matter expert in satellite communications and established the technical architecture known as remote operations. When the whole notion of Predator was initially considered, the CIA made an inquiry of the Air Force as to the technical and personnel requirements to operate it. The result was a very robust package that included over one hundred persons being deployed.

From the CIA perspective, this was completely unacceptable. Our vision ran along the lines of a clandestine operation in a host country where we would minimize the US footprint and signature as a requirement for cooperation from the host country. Putting what we saw as a sprawling Air Force package on the ground was, to pardon the pun, just not going to fly.

That led to the alternative of a forward element to do launch, recovery, and maintenance on the aircraft. The rest of the flight crews, along with all the weather, intel, planning, etc. would be parked in some more permissive country. That site just had to exist within the footprint of the same satellite being used to operate the aircraft. That cut the staffing chart from over a hundred to just seven, maybe ten on the outside. The Agency was happy, the Air Force was happy, and just as important, our host nations were happy. It was one big circle of win.

Albert was the guy who made it all possible. He was able to work the Ku-band satellite communications so that the pilot and sensor operator would take off and set the aircraft into a "racetrack" pattern over the base. At that point, command and control would be transferred via satellite to the remote base, which was then free to prosecute the mission. The launch team was off the clock for ten to twenty hours until the plane came home, and the hand-off reversed. They would land it, refuel and refit, and have it prepped for the next mission.

Alec dropped one of his hidden cards on the table. "There was a big push to pull the plug." His tone sounded a bit like confession. "Honestly, Mark, I had called folks together to start that process, develop a list of shutdown actions." Alec took a breath, as if considering the weight of those words. Then he continued.

"Then Albert had one of his big-idea moments and asked if anybody had thought of moving the remote element, splitting it away from the receiving Ku antenna."

That set my gears turning. We'd need to break the Link Management Assembly in half, then replace inches of copper wire inside the GCS with miles of fiber optic cable outside. I tapped my fingers on the desktop. It was not easy per se, but it was straightforward.

Alec was following the same path. "Yeah, I was good with the sock-puppet explanation, right up to the point that Albert dove into some ATM thing and lost me."

I shrugged, understanding his confusion. I knew I'd have to get Paul Welch to explain the mysteries of asynchronous transfer mode, enough at least to see why we'd need that instead of TCP/IP to get a reliable, real-time link to smoothly control the plane and laser.

Looking out the window I felt a twinge. It took another moment to realize it was hope. Then my practical side kicked in, and I turned to look Alec in the eye. Something told me he had another card as yet unplayed.

"Move it . . . where?"

Alec sighed. "That's the $64,000 question. Since Albert brought it up, he got stuck with developing a handful of options."

I chuffed, imagining Albert's face during the meeting when his well-intentioned question turned into a boomerang. Two brains or no, Albert clearly had forgotten one of the most fundamental rules

THE GOOD, THE BAD, AND THE UGLY

At right: Ahmad Shah Masood, the Lion of Panshir. For twenty-six years, al-Qaeda, the Taliban, the Pakistani ISI, the Soviet KGB and the Afghan communist KHAD had all tried to kill him, unfortunately succeeding on 9/9/2001. His Northern Alliance was an important part of the US strategy before and after 9/11. *Photo courtesy US Department of Defense*

Below left: Mohammed Atef was the chief of al-Qaeda and the de facto right hand of Usama bin Laden. He played a major role in the attack on the USS Cole. His reign of terror was ended by our Predator in November 2001. *Photo courtesy of US Department of Defense*

Below right: Mullah Mohammed Omar was one of the founders of the Taliban, responsible for atrocities like the destruction of the Buddhas of Bamiyan. Predator attacks drove him into hiding; this severely constrained his ability to operate. *Photo courtesy of US Department of Defense*

The first test to see if we could shoot a Hellfire from a Predator, conducted at the Naval Air Weapons Station at China Lake, in California. We went with a ground test because nobody knew if the force of a missile launch would just rip the wing off the aircraft. *Photo courtesy of James "Snake" Clark*

"The pointy end of the stick": the AGM-114K Hellfire missile, with an explosive warhead that weighed just twenty pounds hanging beneath the wing of a Predator. *Photo courtesy of Alec Bierbauer*

The Tantalum sleeve: a brilliant bolt-on modification to the original Hellfire design. On impact, the sleeve would shatter along the diamond-shaped (pearson v-notch) scoring pattern visible across the inside surface. This saturated the area of detonation with white-hot, supersonic fragments that greatly increased lethality against unarmored targets. *Photo courtesy of Alec Bierbauer*

Hellfire missiles with the Tantalum sleeves attached. *Photo courtesy of Alec Bierbauer*

Taco Bell: our test building in a southwest US desert, intended to replicate the dimensions and construction materials of bin Laden's home at Tarnak Farms. The construction process may have overshot the mark just a bit in terms of thickness. *Photo courtesy of James "Snake" Clark*

Success nonetheless. Despite the extremely solid construction, Hellfire proved capable of punching through the heavy wall. Seen from the inside, the missile's explosive force and fragmentation—along with all of the displaced wall material—becomes a cloud of fiery debris that (hopefully) saturates the room, and the intended target. *Photo courtesy of James "Snake" Clark*

It was critical to understand what Hellfire would do against a variety of targets, and what risk it posed to people nearby. Melons served as the poor man's ballistic gel in an SUV penetration test, dressed up with Xerox photos in the name of science. Any resemblance to fearless leaders Snake Clark or "Spoon" was purely coincidental. *Photo courtesy of James "Snake" Clark*

The Delivery Vehicle. Everything to do the job had to be carried by this spindly airframe. You can see the round "MTS ball" camera system under the nose that provided a view in visual as well as thermal. The ball also housed the targeting laser used to guide the Hellfire missile. This was the first unmarked, all-grey Predator. *Photo courtesy of Alec Bierbauer*

Inside the Ground Control Station (GCS), looking right. When you have to cram an entire flight control cockpit along with comms gear inside an intermodal container, things get pretty cramped. *Photo courtesy of Cliff Gross*

Inside the GCS, looking left. The white cowboy hat hung ready for service. *Photo courtesy of Cliff Gross*

The Launch and Recovery workstations. Similar in nature to the GCS, but located at our clandestine base adjacent to the area of operations. While stuck in a remote outpost on the far side of the planet, the L&R team at least enjoyed a little more space when on the stick. Poor payment for a tough and vital job performed so flawlessly. *Photo courtesy of Alec Bierbauer*

Master Sergeant Cliff "Cliffy" Gross, doing his technology magic in a corner of our intel cell in Europe during operations in 2000. *Photo courtesy of Mark Cooter/Cliff Gross*

The Tent, a clever name for our intel and planning cell at the Europe location in 2000. *Photo courtesy of Cliff Gross*

Just one year later, the Tent was replaced by the celebrated "double-wide." The photo shows the double-wide at its non-military location. The GCS sits only feet away, allowing for not only intercom- and phone-support but face-to-face conversations if needed. This adjacency maintained the highest possible level of shared mission awareness across the entire team. *Photo courtesy of James "Snake" Clark*

The GCS, now painted civilian white to blend into its non-military Virginia location. Move along, nothing to see here. *Photo courtesy of James "Snake" Clark*

The BAD (Big-Ass Dish) which was, ah, creatively acquired from Langley Air Force Base under the cover of darkness, then whisked away to our location in Europe. The tech standing on the hub gives you a good sense of scale. *Photo courtesy of Cliff Gross*

The Man in White. This is a still image from the real time video view we had of Usama bin Laden at Tarnak Farms, taken less than nine months after being set on the impossible mission of finding him. Dressed head to toe in white, UBL is in the center of the shot, walking from the truck towards the group to his left and then onward to the mosque in the upper left corner of the image. *Photo courtesy of US Department of Defense*

Tarnak Farms as seen from a Predator wide angle shot. This was a post-strike image after the start of the war and shows the destruction of many of the structures at the al-Qaeda facility. *Photo courtesy of US Department of Defense*

```
Time Zw
12 Sept
 1530        dept   Charleston SC
 2100 z      ARR    China Lake
 3Sept       dept
  00:15      dept   China Lake  CA
  00:45      ARR    Palmdale  CA.
  04:00      Depr   Palmdale  CA.
  09:00      ARR    Andrews  Md.

 14Sept      Dept
 04:45       Dept   Andrews  Md
 12:45       ARR    Ramstein  Ge

 15Sep
 18:30       Depr  Ramstein  GE
             ARR   Tashkent

 16 Sep
 01:15       ARR   Tashkent

    ?        L-100  dept
```

After 9/11, the team acted swiftly to respond to the AQ attack on the United States. This note by Mr. "Snake" Clark served as the Execution Order for the team's movement overseas. As you can see, within hours, the team was moving to engage AQ with little fanfare, guidance, or bureaucracy. Within a day after arrival at the forward location, they launched their first mission well ahead of other US forces. *Photo courtesy of James "Snake" Clark*

The team's challenge coin and motto. While certainly not a household phrase like "Semper Fi" for the Marines or even "Aim High" for the Air Force, it exemplifies the perseverance and dedication of the team and became the foundation for our teamwork and esprit de corps.

It was born out of frustration, particularly when we proved we could get "eyes on UBL" and were told others would engage and eliminate him, but no one came. Time and again we were challenged like this, we would think our mission was doomed to failure, but then a teammate would come up with a solution to overcome the challenge . . . Never Mind We'll Do It Ourselves. *Photo courtesy of Mark Cooter*

Mark (above) and Alec (below) each receiving the National Intelligence Distinguished Service Medal at CIA HQ with Director Tenet and ADCI for Military Support, Lt. Gen. "Soup" Campbell. *Photo courtesy of the Central Intelligence Agency*

Numerous members of the Predator team, representing various component services, gathered at an awards ceremony at CIA HQ in 2003. *Photo courtesy of the Central Intelligence Agency*

A dream reunion: members of the Predator team gathered at the Smithsonian Air and Space Museum in Washington, DC, to say hello to an old friend, the actual "034" Predator. One of the original operational aircraft, '34 survived it all in Afghanistan and now hangs in a place of honor to share her story with generations to come. *Photo courtesy of Michael Marks*

LOST BUT NEVER FORGOTTEN

A real patriot, Jeff Guay tagging the mil-spec version of the GCS with Big Safari colors. RIP Gunny. *Photo courtesy of Cliff Gross*

One of the last photos of Troy Johnson, who we lost in 2015. Troy went out doing what he loved most, flying. Godspeed. *Photo courtesy of the Johnson family*

October 2000. The bodies of those lost in the attack on the USS Cole are unloaded at Ramstein Air Base in Germany, a vivid reminder of the dear price paid for freedom. *Photo courtesy of US Department of Defense*

Back row far right (black jacket): Master Sergeant Donnie Davis, Mark's high school classmate, protecting Karzi before being killed in Afghanistan in 2001. *Photo courtesy of US Army*

THE AUTHORS

Mark Cooter

Mark deployed with the 11 ERS operating the Predator from Bosnia in 1999. After completing a twenty-eight-year AF career, he's continued to do similar work as a defense contractor. Every year he sends a note to the team thanking them for their dedication and teamwork.

Alec Bierbauer

Alec at the Derunta al-Qaeda training camp outside Jalalabad in 2002. Alec continued to prosecute the Global War on Terror with deployments around the globe before transitioning to the corporate world.

Michael Marks

Embedded with US Special Operations in Afghanistan 2011–12. Mike worked shoulder to shoulder with Col. William "Billy" Shaw III (RIP) to significantly enhance America's asymmetric warfare arsenal. He continues to serve the intelligence community.

of a military briefing: if you know enough to ask the question, you know enough to find the answer."

"But since you just asked," Alec said, as if listening to my thoughts, "where would you put it?"

Fuck, I snarled inwardly. The corollary to that basic rule was demonstrating any grasp at all of the problem is likewise volunteering. Scratching my head, I considered the puzzle pieces. The choices were scattered throughout a briar patch of political and practical thorns.

"Well, the antennas can stay where they are; it's the whole MCE that needs to move, ideally to someplace we own outright." I knew as I uttered the words that the devil would once again be in the details. We'd need high physical security, high-bandwidth fiber connectivity, access to high-side and low-side MILnets, uninterruptible power, and, oh yeah, utter invisibility to the outside world.

"Can Albert make this happen all the way back in the US?" I asked.

Hauling several tons of command-center and ground-control equipment back to Terra Americana would doubtlessly fulfill the security and infrastructure requirements. But that sort of move would mean that reliance on "miles" of fiber optic cable just stretched out to some six thousand miles, crossing an ocean floor in the process. A link of that distance might stress physics.

Split ops was one thing; flying an armed aircraft in a combat environment from two points connected within a single satellite footprint attached to a third point via thousands of miles of fiber optic cable was something else. This half-assed notion we were batting about was around the globe. *Extended Split Ops*, no . . . *Long-Distance* . . . the terms didn't begin to touch cabling a signal from the United States across the Atlantic, then banking it off a satellite to some remote patch of sky over Afghanistan. I grumbled under my breath—"remote" felt more like our chances of success.

The light bulb flickered on: *Remote Split Ops.* Well, we'd have a name for it anyway.

"Okay," I resumed, picking up steam. Half-baked ideas and hard work had gotten us this far; there was no point in quitting now. "We've got the obvious choices to start with, the ones with the most robust and redundant communication links." My mind ran down the usual suspects. "We could set up at Fort Belvoir, maybe Langley Air Force Base, or, God help us, the Pentagon. DOD owns the ground; we could keep a tight lid on security and pretty much call our own shots with facilities."

Alec seemed to process the suggestions thoughtfully, then offered up an unexpected option. "What if we reverse the whole layout, run the whole thing on a ship downrange? We have persistent assets in the Indian Ocean, maybe something in the Med."

I didn't like the sound of that one, not the least of which was having little desire to spend the next year of my life as a bobber on a big ocean. I had joined the Air Force and not the Navy for a reason.

"Sounds sketchy," I said. "A ship can't drag fiber, so we'd have another wireless connection to contend with. That creates bandwidth issues and a whole lot of ways to lose link. I'm guessing uptime would be far more weather-dependent than a land-based op. Logistics and maintenance lines would be far more complicated—"

"Got it," Alec said with a dismissive tone. "No ships."

Whew! I breathed a sigh of relief. I didn't want to have to tell Ginger we were going to be on a ship. That would be worse than dealing with a porta-potty.

Alec's next statement had that sort of casual just-had-a-thought tempo that struck me as the place he was headed all along. With a hint of repressed smile, he asked, "What would you think about setting up shop back here in the States?"

CHANGE CAN BE PAINFUL
July 2001

"Mark, your leadership is setting you up for indictment."

It didn't help my sudden mood swing that the assessment was delivered face to face by an Air Force brigadier general who pulled me out of a meeting to say so. He was the deputy director for operations on the Joint Staff, so his opinion had some weight behind it.

The shock was considerable. I flashed back to 1988 when, as a second lieutenant, I watched Oliver North get indicted on sixteen felony counts and convicted of three. In the blink of an eye he'd gone from Marine wunderkind to a pariah neck-deep in the Iran-Contra affair, in a cloud of allegations that included drug dealing, accepting bribes, and destruction of documents. His military career evaporated overnight.

Our track record of, well, creative problem-solving flashed to mind, leading to mental pictures of Alec and I sharing a cell in Leavenworth. In a brisk lap through the office I waved off Alec's opening question, filling the sudden silence with a growled "I need to go see General Campbell."

Lieutenant General John "Soup" Campbell was the associate director of central intelligence for military support. In layman's terms that meant he was Director Tenet's principal adviser on all military matters, the top military officer at the CIA. It didn't hurt that he was Air Force. I related the incident and asked for his guidance. Campbell lived up to his reputation for leadership; he allayed my fears and assured me that the mission was critical, and he had our backs. I went back to the office, comfortable that any ill-intent aimed in our direction would have to get through Campbell.

My bulwark restored, I headed back to check in with Alec. That led to a deeper exploration of setting up a Remote Split Ops command center.

The choices were almost immediately overwhelming. We didn't need a big structure; for the most part, any sort of building would do. We needed some office space, a couple meeting rooms, and, oh yeah, bathrooms—real bathrooms.

It struck me that the idea of hide-in-plain sight wasn't without precedent. Virginia has been home to all sorts of secret government facilities. For the longest, time Mount Weather was home to what started out as the innocuously named Civilian Public Service facility but grew to be a complex underground facility capable of housing the collections of the National Gallery of Art, along with members of Congress, in the event of nuclear war. You'd never have known it driving through the sleepy little town of Purcellville.

In terms of James Bond movie cool, the Greenbriar put Mount Weather to shame. A luxurious golf resort and spa by day, the Greenbriar concealed huge secret walls that could swing open to reveal multiton steel blast doors leading to a reinforced concrete vault capable of sustaining the US government in times of nuclear war or other world-ending catastrophe.

It didn't seem likely that we'd end up with underground vaults or hidden tunnels; our choices leaned more toward the mundane. Abandoned or underutilized commercial buildings topped the list, starting with industrial facilities that offered a warehouse-like interior.

With little more than hope and a map, we set off in two cars. Alec and I took lead with Alec's boss Cofer Black and his entourage following in a second. Our first stop was an old print shop, which briefly looked promising, but the landlords seemed far too interested in what we planned on doing. Alec bristled at the intrusion. We looked at another option, but it would have put us uncomfortably close to a daycare center, a prospect that seemed fraught with disaster.

Technically, a residential property wasn't out of the question; it just needed to be large enough and, more important, very private.

Virginia is home to numerous mansion properties, some dating back as far as the Civil War. Many were still in family hands, passed down through generations. Others had been converted to exquisite B&Bs, like the Ruffner House out in Luray. Sadly, many of them had fallen into disrepair, better suited to housing bats than birds of prey. With a classified ads splattered with red marker scratch-outs between us, Alec and I rounded the curve of a long, tree-lined driveway. I felt my jaw go just a little slack. Bingo!

The house was a magnificent turn of the century wood-frame residence, a sprawling white two-story building with end-to-end windows stacked all the way up to the dormered roofline. This piece of history would be a massive upgrade from our prior life in a leaky tent on an airfield. Like any competent Air Force officer, I quickly scoped out a perfect spot in the house for a bar, should uh . . . should we ever be called on to entertain VIPs.

Driving away, I was already neck-deep into allocating floor-space when Alec braked to a stop in what amounted to nothing more than a gap in the woods. I glanced around, puzzled. Best I could figure, Alec was pulling over to pee in the bushes.

"What do you think?" he asked.

"Aw, the house is perfect," I replied, "out of sight but very comfy and—" My commentary trailed off when I realized that his question was not directed at the house but at something out the window. I squinted, seeing nothing but what looked like a clearing tucked back in the trees. Confusion smeared my mental floorplan.

"What, out here in the middle of fucking nowhere?" I was looking for the joke to end.

"Sure, why not?"

I climbed out of the van, still confused, when I realized my foot was planted on flat asphalt. We weren't entirely nowhere; we were on the only paved spot in nowhere. *Well, that's so much better.*

I scanned around, seeing little more than potholes. By the piles of crap off to one side, it had been relegated to a dumping ground.

Alec paced the barren pavement as if in keen appraisal. He had the look of someone imagining what might be—I was fixated on what was missing—no plumbing, no electrical, not even a stub poking up out of the ground that might have offered a hope of networking. In the back of my mind I could already see the porta-potty off to one side and Ginger hunting me with a knife.

I went back over our priority list as Alec continued his survey. I still would have preferred the familiar comfort of an Air Force base—but for God's sake, not out here in the freakin' woods.

Still, I caught myself thinking, *We couldn't beat it for privacy.* Woods surrounded the area, and based on the reasonable proximity of other buildings, the hurdles of hauling power and internet out there didn't seem insurmountable. Hell, I could drag cables across the ground if I had to. Against all common sense, I caught myself muttering under my breath, "We drop a mobile home right here, plenty of room for a comm shed and a couple of GCS boxes. . . . It'd work."

The location had decent proximity to a number of critical infrastructures, ranging from airports to metro stations. It was a pocket of "far away" that wasn't all that far.

Members of Cofer's entourage didn't share my limited enthusiasm. Shit may roll downhill, but doubt quickly floats up to the top, and Cofer twigged on the conflict. He cut to the chase.

"So where do you want to put your command center, major?" It carried a tone that demanded decision.

I was tired of tromping my ass from one bad choice to the next, and it was pretty clear I wasn't getting my mansion bar. Fuck it. "Right here, sir."

Alec shot me a quick glance that asked, *You sure about this?* I just nodded in return.

With a curt nod Cofer said, "Me too," and climbed back into his car. In a moment it rolled off in a cloud of dust, leaving Alec and I standing in our clearing.

I sighed, feeling my shoulders sag. The last glimmering image of the Predator bar hung before me like a distant mirage before fading away forever. But one thing was clear: we had a home.

After a rapid lap with pen and paper sketching out locations for the major pieces, we set off to engage a couple of contractors, Pete and Martin, to assist Paul and Cliffy to get to work on a double-wide. With a wistful parting nod to the mansion bar, I set off to move the MCE.

The first step was a call I didn't look forward to making. After months of operating under half-baked, duct-tape-and-porta-potty conditions, Ginger had been at her veritable ribbon-cutting moment to declare the site was finally up to snuff when I told her to put everything on hold in June. I was thankful to be calling from stateside when I dropped the second shoe.

"Tear it all down, Ginger. We're shipping it to Virginia."

Ginger was colorful in her response, her Kentucky accent stronger as she picked up steam. Once she had it out of her system, she checked back in as an amazing intelligence officer and professional. Like the rest of us, she knew the silver lining in this call was the fact that we were spinning back up to deploy. Regardless of the political statements from our bosses, we knew there had been a sizeable chance that things would have just been shut down. This move was a good indicator we were moving forward, even if our pilots and gear were moving six thousand miles further from their target.

THE HOUSE THAT PAUL BUILT
August 2001

As we expected, communications proved to be one of the most demanding demons to wrangle. When we stretch a line thousands of miles, on top of everything else we need to cope with the sheer weight of distance. Terms like lag and ping time rarely have any impact on life except for competitive video gamers, because the actual transit time from a home or office to the internet is measured in dozens or, rarely, hundreds of miles. When that number hits the thousands, all those fractions of seconds add up.

So do echoes and dropouts, signal noise, and degradation—all the little anomalies that conspire to turn a clean, clear signal into a meaningless static. No one has seen the Verizon guy asking, "Can you hear me now?" from a mountain in Afghanistan.

If a solution was to be found, we had to call in our best. "Alec, whatever it takes, you need to pay for Paul Welch to come here and take a look at these comm issues." I got no argument; Alec knew I was right. We needed Paul ASAP.

The double-wide was already in place, sitting on stacks of cinderblocks. It was a long stretch of tan with dark wood-colored trim. At the moment it was surrounded by ladders and sawhorses, hemorrhaging wires from a dozen points along its exterior.

Throughout the installation process, Paul played a huge role in our race for operational readiness. Obstacles appearing out of left field were more the rule than the exception, a persistent hazard when trying to build something that had never before been attempted. Since we were inventing new problems, we had to come up with new ways to troubleshoot them. Paul and Cliffy cobbled together a system that could replicate real-world performance of communications from the GCS all the way to a tactical operations center sitting in some desolate spot in a theater of war. This gave them a simulator of sorts, against which to identify and ultimately slay a vast number of communication demons. Those results were turned into protocols that could be invaluable if a real-world demon raised its head in the midst of a crisis.

But every newfangled problem dragged a couple of old ones along for the ride, most notably the immutable power of a low-level admin to throw a wrench into the gears and derail a national priority. Few things are as daunting as a bureaucrat vested with the power to make decisions on subjects they don't understand.

In one notable case it was an IT tech, someone responsible for maintaining the operating systems and policing software on the secure network. The interagency nature of this collaboration doubled down on the typical challenges faced when companies want to collaborate. The Air Force ran on one system, while the Agency ran on a different system altogether; both were on the "high side"—top secret—but unable to talk to one another. It was going to be an utter bitch to run a real-time op if the top two guys couldn't move target data between each other. Either we had to bridge that gap or come up with our own top-secret network.

The Air Force side of my brain needed to correlate threat intel from both the CIA and NSA with maps and flight data normally contained in an Air Force program called FalconView. But the CIA was apparently not going to allow an "outside" piece of software on its network. It was the endless battle, Agency versus DOD, ops versus intel. We just needed shit to work. The "why" was way, way above their pay grade. I pushed, they pushed back. I give 'em credit for a dogged tenacity to blindly follow the rules; their answer no quickly became hell no.

Paul could see the stress this was causing and waved me off, saying "It's all under control." Having an overflowing plate of my own, I was thankful for the handoff. Paul rounded up Cliffy, Pete, and Martin and set out to build our own high-side network.

Not all of this was a matter of wires and bit rates. Every network comes with its own set of permissions, and we needed to create a new class of user, the bastard stepchild of DOD and Agency heritage. To make that work we needed a sacrificial lamb, and all eyes settled on me. Alec's security officer Linda Anderson, a bona-fide miracle worker, did some bizarre, behind-the-curtains magic and turned me into a patchwork of access privileges, complete with a homemade badge. Somehow, beyond all logic, the whole thing flew.

The next thing I knew, I had a working terminal, connected to Alec, where we could successfully work threat and targeting data in real time. In addition, we had a network of computers for our team to forecast weather, analyze intelligence, and plan and execute operations. It wasn't fancy, but it worked. I marveled at a course of action I never had thought of, much less knew how to begin to execute. But I liked the logic: sometimes it's easier to go around a problem than to plow through it. That was good, as the time had come to switch gears, put on my operations-officer hat, and meet some old friends back at China Lake.

WE'RE GETTING THE BAND BACK TOGETHER
August 2001

Alec and I walked into the old worn-out military building at China Lake and climbed up the stairs to the second floor. The conference room was filled with a group of hand-picked pilots and sensor operators. Some I knew from Indian Springs; others had been chosen on a best-in-class reputation. I nodded to those I knew, reading "what the hell is Cooter doing here, and who's the clean-cut dude with him" on their expressions. The look on the remaining faces, the ones I didn't know, gave off something more like "who the fuck are these guys?"

I unlocked and unzipped the blue bag that I always used to carry my classified program information. The room was quiet as I cut straight to the chase.

"I'm here to see if you're interested in joining me on a special project. Before I can discuss the program, I need each of you to sign a nondisclosure agreement."

The words sounded like dialogue from a Hollywood spy thriller, but the group of men and women in front of me weren't actors playing movie roles. They were the real deal.

I pulled a stack of forms out of the bag and laid them on the table. The document was no joke. People who carry a security clearance are used to signing an endless stream of waivers and acknowledgements regarding the value of national security information and the horrific price to pay for divulging it. This one was worse than most. The word "thumbscrews" did not appear per se, but the implications were clear.

"If you're not interested, please leave now." I watched the group as I spoke, looking for a flinch.

In a split second, all those that knew me reached for a form with a smile on their face, pens out. A heartbeat later, those that didn't know me followed suit. No one left the room. It is humbling as a leader to feel the trust of your team when you watch them run off a cliff on blind trust, flipping pages and scribbling signatures.

The formalities concluded and the forms stuffed back in the blue bag, I pulled out a map of Afghanistan. Atop that I slapped a picture of Tarnak Farm and a frame from the Man in White video.

"We hunted bin Laden in 2000. Now we're going back . . . only this time we're going armed."

I saw a mix of wordless responses: eyes flashed wide, bodies leaned forward, and a great many grins that looked like wolves watching the sheep. Only one question was raised: "When do we start?"

Darran was the first to speak, which came as no surprise to those who knew him. An Air Force major, Darran was a special ops-qualified C-141 navigator, a Predator pilot, and my first choice as opposite-shift mission commander. We were at Indian Springs together, and he had my utmost confidence. Darran was a no-nonsense guy, never shy about calling bullshit on behalf of his own interests and those who looked up to him. He came right out with the hard questions, the elephants in the room that I'd bet nobody expected to be addressed.

Dispensing with the political tap-dance, I answered, "Right now." I wrapped with a blunt stare that asked, *So what do you think of that, smart guy?*

True to his character, Darran offered his endorsement, acceptance, and profound words of wisdom with a single monotone word: "Cool."

With our returning all-stars rounded out with a crop of promising free agents, our fantasy team headed straight into a series of heavy workouts. As big-boy fun as it might sound to shoot missiles at buildings day in and day out, a whole lot of muscle-straining, brain-cramping effort went into that click of the trigger. Even our missile, like an athlete pushed to the limit, was sent off to the proverbial ice bath for a long cold soak.

A cold-soak mission profile was a twenty-hour, full-on dress rehearsal: conducted at altitude, replicating not only the extended time of a real mission, but of staying aloft in that miserable cold for the duration. It was one thing to take a Hellfire up to twenty thousand feet and shoot, but another to let every centimeter of it hang in the freezing cold for hours on end before we expect it to work. In that kind of environment, things can gum up, and fragile materials can crack. When we add the ceaseless vibration and the bucking about as a light plane rides rough skies, we can shake the dental fillings out of some pretty robust equipment.

The range wasn't happy to have to stay open all night. They were even less happy that some of our guys were running around in the dark providing something to watch. Without that, staring at the moon-lit desert floor of China Lake would have been the definition of forced boredom. But although these missions were painful, they were important.

Missions like the cold soak also test the human links in the chain. Twenty hours is a long time to stare at a battery of screens and gauges and to get by on little food, less sleep, and short sprints to the bathroom. Attention can slip; so can senses of humor. All of our new crews participated in the tests, so the new pilots got a chance to learn the Remote Spilt Ops procedures, and the new sensor operators got trained on the MTS ball for the first time, though none got to actually fire a missile.

It gave us time to talk a lot about Afghanistan and our TTPs,[64] to work through finite adjustments. In the end, the flights went off without a hitch, the crews gelled perfectly, there were no cold-soak issues, and we destroyed our intended target. Our weapons, and the hands to wield them, were ready.

THE MONKEY SWITCH
August–September 2001

While the weapon platform was coming together, up and down the food chain we continued with the wargaming, and in some cases handwringing. Not wanting a repeat of the toothless tiger from September 2000, we had to be sure that as a nation we could stomach actually shooting at the target when we found him.

Thus far we had resolved a fairly defined set of ROE that dictated who we could shoot and under what conditions. An unexpected outcome of the tabletop exercise was a challenge from the OSD lawyer that a military member couldn't be the one to pull the trigger. That might strike the layman as an artificial or meaningless

---

64 Tactics, Techniques, and Procedures

distinction—anybody who put a bullet in Adolf Hitler's melon would have been hailed a hero, right?

But this was a different age, and if recent history had taught us anything it's that for every action there is an equal and opposite litigation. A soldier shooting a suicide bomber is an act of national defense, but a civilian doing the same thing might find himself charged with murder.

Set aside all the horror scenarios where a missile runs amok and hits a hospital or orphanage; those are career-enders any way you look at it. But focus instead for just a moment on the hypothetical perfect mission, where you drop one round on the head of an approved target and a year later a team of lawyers determines that you didn't have the proper authority. Concepts like right and wrong have little impact in the modern courtroom. We were going to need explicit guidance on the subject before anybody touched a trigger in Virginia that fired a Hellfire at a target in an undeclared hunt for America's most wanted terrorist in Afghanistan.

On a normal mission, a number of authorities were represented in the GCS. We had Air Force representing sworn combatants, CIA personnel, as well as a number of civilian contractors. Although one of us might be the right guy or gal, it was pretty certain that a lot of us were disqualified.

The legal beagles determined that the authority to exercise lethal force in this context resided with the CIA, so it would need to be a CIA officer who released the weapon. But the pilots flying the plane were Air Force personnel, and the CIA had no meaningful capacity to train up alternatives. It was even less practical to attempt a Chinese fire drill and swap seats at the moment of truth.

Eventually, a solution was proposed that satisfied a majority of the lawyers: a separate switch that, once enabled by the pilot, would actually cause the missile to fire. We tasked Big Safari to wire a toggle switch with a red safety cover directly to the control panel between the pilot and the sensor operator of the GCS. After the Air Force guys flew the plane and aimed the laser, the ranking CIA officer in the room would flip up the safety cover and, on the mark, flip the switch.

When I first saw the new switch installed, I had another Tom Clancy flashback, this time from a scene in *The Hunt for Red October*. Onboard a ship in the Atlantic, CIA bigwig James Earl Jones flashes his ID, presses the red button on a cluttered console, and detonates a torpedo before it impacted a submarine. He said, "I was never here" and disappears. The image brought a flush of concern—we'd have to make sure that an Agency guy touching our switch didn't somehow cause the Hellfire to spontaneously self-destruct.

Of course, nobody in the military universe can touch a gizmo without training and formal certification to use it. So, Alec, Rich Blee, and Alec's alternate, Brian, went through a class on the proper way to flip a plastic cover, count down from ten, and toggle a switch. If you just read the previous sentence, congratulations. You are now qualified as well.

Having three "certified" switch-flippers would ensure that at least one of them would be available during potential engagement windows. With repeated demonstrations of competence, the training took about thirty minutes. In lieu of a certificate, I awarded each student a banana, commensurate with the level of skill required. As a result, the system became known as the monkey switch.

It was oft suggested that the monkey switch was just a meaningless prop that had been super-glued to the console, with nary a wire behind it, as a fiction designed to appease the lawyers. From my own perspective, no Airman in his right mind would hand an Agency guy the trigger to fire a live missile. The reality of the monkey switch will go with me to my grave.

Just before Labor Day 2001, the Boeing C-17 touched down at Andrews Air Force Base in Prince George's County, Maryland, just a stone's throw outside of Washington, DC. I watched the massive cargo ramp settle down to the tarmac with a hydraulic whine. A big steel conex box sat in the back, unmistakably clad in swirls of desert camouflage.

Leaning against the GCS were Cliffy and Staff Sergeant Chris J., the latter one of our superstar analysts from 2000. Cliffy trotted down the ramp as cargo handlers began to disconnect the tie-downs that kept the GCS in place during six thousand miles of flight.

I shook Cliffy's hand and threw a glance at the GCS. "How's she looking?"

Cliffy shot back a confident nod. "All good, major. She's packed tight and should be game-ready with minimal set-up."

I watched as ground equipment hefted the GCS and began to back it down the ramp. In a few minutes the metal box sat on the tarmac, and I took a moment to look inside.

Everything was in place. Cable clusters were tied off and marked, every moveable component lashed down. As service members who lived on their mobility, Air Force guys knew how to pack and stow like nobody else. Cliffy leaned against the door frame and tipped his head toward a curve of white in the corner. I had to smile; even my cowboy hat made the trip in place.

With a wordless nod, I patted Cliffy on the shoulder. He had knocked the relocation effort out of the park, cutting through red tape and mechanical challenges to get an entire aircraft control center across the Atlantic is less than two weeks.

Despite my rank, we were respectfully shooshed out by Airmen anxious to get this oversized box onto a truck and off their tarmac. We got out of the way and watched as it was loaded onto a flatbed destined for a nameless warehouse to the south in the Virginia suburbs.

As it drove off, I looked at the multicolor paint job that screamed "military" and thought to myself, *That's the first thing that's gotta go*. One of the requirements to keeping a low profile, in addition to having our GCS and double-wide in the woods, was to remove all camouflage and make sure nobody came to work or visit wearing a uniform.

Once at the warehouse, eradication of the telltale camo was hardly work for Michelangelo. The task instead fell to me; Paul Welch, who had arrived again from Germany on August 29; Cliffy; Linda, our trusted security officer; along with a logistics guy. Labor Day weekend found us clambering on and around the GCS, armed with everything from Wagner power painters to hand-rollers, slathering white paint on every inch of brown.

Several hours later we stood back, every one of us speckled in white dots like some sort of aberrant version of the whitewashing boys from Tom Sawyer. Only what sat before us was no picket fence, but an altogether hideous box that looked, if anything, like a decommissioned ice cream hauler. But nobody in his or her right mind would take it for a top-secret command center. As far as I was concerned, that made it beautiful.

Declaring our lack of artistic ability to actually be a cunning use of feigned sloppiness as part of the deception, the GCS was trucked over to the pad in the woods where Paul and Cliffy began the task of coordinating the trenching and installation of power lines, plumbing, and networking. Linda promptly swapped her painter's hat for a security hat to diligently regulate access control for the countless military, contractors, and visitors who would need to be processed through to our oasis in the woods.

Things were coming together quickly, but in the back of my mind the clock was ticking ever faster. Somewhere out there, bin Laden likely felt that his plans were moving along as well.

Adding Hellfire to the program had proven our capability to hit a target and deliver one hell of a punch in the process. We were quickly moving to prove our adopted motto to "do it ourselves."

As I ran through the checklist, I figured we were just about as ready as we could be. All we needed was the green light to proceed. Of course, that just might prove to be the one hurdle we couldn't cheat our way over.

Within a week we had uncocooned the GCS and routed cables across the asphalt pad into the double-wide. Rubber speedbumps, strips of heavy black and yellow were in place to protect the wires from an errant foot or car. Though I had already wrung far more miracles than I deserved from Paul and Cliffy, I needed another wave of the magic wand.

"What does it take to splice the ARC-210 radio from the GCS into the combined networks of the NSA and CIA?" I asked, scratching my head. The acronym-laden question was hard enough to pronounce, much less answer. Those networks were designed to keep people out.

Cliffy pondered the question for a moment, then said, "Give me a few." He grabbed his two-liter bottle of Diet Coke, took a drag on a smoke, and headed to the GCS with Paul. In what seemed to be no time at all they passed by when, as casual as an afterthought, Cliffy said, "Oh, your network hook-up is running."

I looked at the two grinning faces, my eyes snapping between them and the GCS and back. Taking the comment as a tease on my Mission Impossible request, I waited for one of them to break and let me in on the joke. All I got was two flat stares, as if they had told me the sky was blue.

"It's working? But the NSA . . . OK, *how*?" I stammered, incredulous.

Cliffy shrugged, clearly unimpressed, and said, "Twelve-dollar part from Radio Shack."

With barely a blink, he turned and walked off to his next task as though he'd just connected a cable TV box. I fought the grin, thankful as hell for an absolutely vital cross-network capability. Having hacked the CIA and NSA at the same time, Cliffy and Paul strode off in search of a worthier challenge.

Shaking my head, the grin won. Damn, it was good to work with the renegades.

The last act was less magic and more inspired innovation. We had a desperate need to put more information into the hands of the pilots and sensor operators.

Cliffy came up with the idea to Velcro additional flatscreens in the GCS. Something of a digital kneeboard, the screen displayed a zoomable scrolling map with overlays of threats, reference imagery, and weather data. It created a graphic interface between the team members in the double-wide, a means for them to push things like satellite imagery, annotated FalconView maps, and navigation plots. The upshot was like a group Vulcan mind meld that gave the flight crew in the GCS instant, nonintrusive access to the experts parked just a few feet away in the double-wide. This change created a whole new synergy that proved invaluable to flight ops.

Despite the low-budget housing, the double-wide was every bit a working command center. Most of the middle of the trailer was a single large room ringed with long tables. We had stations for weather, targeting, threat analysis, and imagery analysis to build products for the sensor operators. Back and left was a room packed with critical spare parts and a folding cot for the inevitable double or triple shifts. It was a considerable relief that the trailer came with a working toilet. Ginger would be happy.

Despite the added jump across the Atlantic Ocean, we had a better handle on comms here than anything we'd had up to this point. That is not to suggest things were pretty by any stretch—runs of cables were zip-tied to the ceiling, and ad hoc was everywhere.

We were just over a week into September when I stood near the edge of our little compound, amazed at the work our team had done in so short a time, under the numerous constraints imposed by working in one of the world's most secret environments. The heat of Virginia summer was just beginning to diminish. Pretty soon the dense canopy of leaves overhead would blaze with yellow and orange hues. I'd stack Virginia in the fall up against just about any place on earth in terms of natural beauty. With luck, we would find some quiet time to enjoy it.

12: UNTHINKABLE

ALEC BIERBAUER

THE DAY THAT CHANGED EVERYTHING
September 11, 2001

The face in the bathroom mirror stared back at me with blood-shot eyes. Forty-eight hours with barely a hint of sleep had me looking ragged around the edges, a long forty-eight hours since Ahmad Shah Massoud had been murdered and forty-five since his corpse had been stuffed inside a refrigerator.

Topping the list of worst-kept secrets would have to be the US government's relationship with the Afghan Northern Alliance, led by the Lion of the Panjshir, Ahmad Shah Massoud. Let there be no mistake, Massoud was a warlord like many before him, a thug with a highlight reel that spanned human rights violations and drug dealing. But Massoud had a brilliant tactical mind and thus far had been our most valuable resource for countering the Taliban in Afghanistan. Lacking a perfect solution, we had settled for a realistic one.

Just shy of fifty years old, Massoud was a powerful military commander who had earned his stripes fighting the Soviets in the 1970s and 1980s. Scarred, grizzled, perhaps a little larger than life, Massoud commanded the last remaining force that was holding off the Taliban from owning all of Afghanistan. Until two days ago, he had been our best hope on the ground.

Massoud had been a close associate of the CIA, dating back to the proxy war against the Soviets. It was safe to say he was responsible for killing more Russians than any other warlord in Afghanistan. Part of his success hung on battlefield savvy; the other part was his control of key terrain. He had denied the Panjshir Valley to the Taliban with the same ruthless efficiency that he dispensed to the Russians. Over the span of decades, every effort to carve a path into the mountains was driven back.

We had no false assumptions about controlling Massoud in the classic game of case officer and asset. But we had a strong liaison relationship, a shared set of interests perhaps. Whatever the label, it was the closest thing we had to ties with a legitimate government in Afghanistan. Almost single-handedly, Massoud was responsible for preventing the Taliban from finalizing control of all Afghanistan and turning their attentions elsewhere.

The US government bet heavily on Massoud. But the Lion was willing to eat friend and foe alike if it advanced his agenda and routinely pushed back against the United States as much as he bashed against the Taliban.

It was no stretch to believe Massoud and his insurgent force could have brought violence to bear on bin Laden if they chose to. But Massoud was smart enough to know that if he gave the United States what it wanted, it would quickly lose interest in Afghanistan and move on to the next five-meter target (the next most important target, in other words). After all, that's precisely what we did as soon as the Russians left. Any hope he had of garnering our support hinged on dangling a six-foot piece of meat just beyond our reach.

The Taliban was Massoud's primary target; bin Laden was not. To be sure, we repeatedly made a point of highlighting the support that bin Laden poured out to the Taliban. He was well aware of the fact; Massoud knew that he was something of a training target for bin Laden's junior al-Qaeda. There were numerous cases of AQ trainees being sent into fight against Massoud as a means to prove their mettle in combat.

The arguments weren't lost on Massoud, he just had higher priorities. The Taliban was his enemy, plain and simple.

He countered by highlighting that the United States "ally" in Pakistan—a key enemy of Massoud's—was providing substantial support to bin Laden in the form of intelligence as well as funds and equipment. With a change in our position on Pakistan however . . . Once again, he dangled the piece of meat.

The net effect was that neither side was willing to act as a proxy for the other. Massoud would be concerned with bin Laden only so far as it impacted his interests, and the United States was unwilling to alter the delicate balance it had with the Pakistanis in favor of Massoud.

Throughout the late 1990s and into 2000, Massoud and several of his key officers were in a regular dialogue with CIA officers in both their Panjshir Valley frontline headquarters and at various meeting locations in Central Asia, Europe, and the United States. I had been very involved with events in Afghanistan since the mid-90s.

I frequently had to facilitate sensitive dialogue with our liaison partners, so much so that I even had an Inmarsat antenna and phone concealed in my suburban Maryland home. During the rare time off, I would still lug the Inmarsat package on hikes along the Appalachian Trail.

I was home early on September 9, thankfully not up in the mountains, when the shrill ring of the secure phone jolted me out of bed. I blinked, focusing on the red digits that read 2:00 a.m. as I fumbled with the handset. It was my regular contact, but this was no regular phone call. He was screaming and crying that the "great leader is dead."

Massoud had been killed by two al-Qaeda operatives posing as television reporters. They concealed a bomb in a television camera and detonated it in the middle of an interview. The effects were catastrophic.

While the first "journalist" was shredded by the bomb, the second had survived to make a break from the interview room. He was gunned down by one of Massoud's guards just steps outside the building.

Admittedly, a refrigerator was no proper conveyance for a man of Massoud's standing. But our contact expressed grave concerns regarding the fragile state of the Northern Alliance, glued together by the charisma and reputation of Massoud. This bombing created the awful potential for an all-out Taliban offensive into the Panjshir Valley if news of Massoud's death became public. At the moment only a handful of people knew the great Lion had taken his last breath. Their immediate action was to stuff him in the icebox and whisk him away to Tajikistan, claiming that he was injured but recovering in a private hospital. If only in spirit, Massoud was still in control of the Northern Alliance. This was a great bit of deception that required very little support from the CIA, but it wouldn't hold for long. When the truth broke, everything would come unglued.

What followed was a hurried drive to headquarters and a marathon session of planning to conceal his death until we could shore up the stability of the Northern Alliance. They would need a new leader, but there was no clear line of succession. In-fighting for the job would take a proverbial axe to the tenuous strings of allegiance in an Afghanistan defined by tribalism and ethnic division. Theirs is a rich history of people turning on one another for the slightest perceived difference.

The morning of September 11 found me surly and a bit disheveled. For the last forty-eight hours we had worked to preserve our precarious foothold in Afghanistan. After a lap through the washroom to put on my best game face, I walked to Cofer's office. A big flat-screen TV hung on the wall in the waiting room. Al Roker stood in front of a crowd predicting a perfect fall morning.

He must be in New York, I thought glumly, *because a shitstorm is brewing here in northern Virginia.*

I walked into Cofer's inner office and delivered a less-than-stellar update on our efforts. I laid out what we had done, what we were trying to do next, and why Massoud's body was in a fridge. The meeting was short, factual, and devoid of any cheer. I would have said it wasn't my worst briefing experience, right up to the point that an autograph book flew off his desk. Having been highly trained by the CIA in the subtle art of reading people, I took that as my cue to leave.

As I hustled out of the office I saw Libby, Cofer's office manager and omniscient controller of all things in the front office. She told me to look at the TV. Something in her tone stopped me in my tracks, and I turned quickly.

The screen no longer carried Al Roker's grin or the sight of a perfect fall morning. Amid the background patter of people crying, "Oh my God," a column of black smoke poured out the side of a skyscraper. Some news anchor was on the phone with an eyewitness, trying to confirm a sketchy report that a plane had just hit the World Trade Center.

The sight of the burning building sucked the air out of my lungs. The news feed cut to a close-up, and I saw an enormous, oblong hole in the face of the building, with a series of smaller holes tattering the adjacent side. The holes lined up—terms like "entry wound" and "exit wound" crawled up the back of my mind.

The entry wound was huge, but not the sort of round ragged crater you get from a bomb detonation. The horizontal gash looked more like a stab wound from a wide blade, hemorrhaging smoke and fire. A billion bits of paper filled the air, fluttering down like snow.

My mind grappled to put a scale on the image. The towers were just over a hundred stories tall, but how wide? It had to be less than half a city block, call it a couple hundred feet at the most. The entry wound looked to span most of it. That put the hole past the wing-span of a stray Cessna. A gut-sick feeling clawed its way into my chest—that hole was a whole lot bigger than a Cessna. Hell, it was probably bigger than the hundred and thirty-odd foot wingspan of a C-130. Unless my math was way off, the tower had been hit by some kind of commercial airliner.

That dawning shoved the intelligence officer part of my brain in a lurch from the "what" to the "why." I had enough experience piloting small aircraft to know that there was no way some off-course airliner would plow into something as big as the twin tow-ers. I glanced at Libby, her face ashen as she stared at the screen, then I took off for the GRC.

From Cofer's office, the GRC was normally a good five-minute walk—I covered it in three. It took just a moment to get through the cypher-lock steel door, a necessity as the unassuming address was home to the inner sanctum of the CIA's crisis communications for the Counterterrorism Center.

For all the things Hollywood exaggerates, the GRC was pretty much what you would expect to see in a big-budget spy movie. At present, the main screen was filled with an oversized image of the smoking hole in the side of the World Trade Center.

Hal was there, as was Mark, who gave Ginger and Paul some quick instructions, and they hastily departed. We barely had a chance to exchange a harried WTF as I followed their gaze at the largest main screen. The huge high-def TV gave us the eerie impres-sion of looking out the window of an adjacent building as a second aircraft, undeniably a commercial airliner, slammed into the second World Trade Center tower. Whatever had crawled from my gut up into my chest simply exploded, taking my heart and breath with it.

The human part of me felt suddenly heavy—some awful mix of tired and old and badly injured. That part of me wanted to sit down, to give my brain a minute to sift for sense in the impossibility my eyes were reporting. But the image on the screen gave rise to a hundred unanswered questions, and the CIA guy in me stepped up to fill the void. We'd undoubtedly need to collect a million bits of intel before we could piece the whole story together. But as the second fireball rolled up the side of the tower and dissolved into a cloud of oil-black smoke, one thing was clear: America was now at war. And I had a damn good idea with whom.

The shock, the sickness, the immenseness of it all, lasted only a heartbeat. The very natural human reflex to gape in slack-jawed horror was overrun by my pressing need to act. Aside from emergency-response personnel who would be grabbing gear and rushing toward the disaster, we were one of the few groups in the country that had something of dire urgency to do this very moment.

The phone circuits came up busy. Without a word, Mark bolted out through the door. I knew he needed to get word to his family that he was okay, then head full-throttle for the double-wide. As hard as it was, I tore my eyes off the screen and struggled to work the problem.

How many compromised planes are still in the air? Targets? Shit, too many possibles to count. The White House, the Capitol— all recognizable from the air.

That last horrific prospect was validated just after 9:30 when reports came of an event at the Pentagon. An explosion, maybe a crash—the details were sketchy. I stopped dead in my tracks, gut-punched again by the first shaky images of the massive concrete fortress, black smoke rising above the Virginia skyline. Unlike the World Trade Center, I had friends in that building, people I loved and respected. Hell, that was Mark's normal workplace. He doubtlessly had more friends in there than I did. Now the entire face of one wedge looked like it had been hit with a flaming sledgehammer. But which wedge?

I snatched a phone and began punching a line to the Pentagon when computer screens all around me lit up with a blaze-orange banner. Under orders of the director of intelligence, the CIA was being evacuated.

I stood flat-footed and blinked. Evac? The word didn't make sense, not now, not here. Yeah sure, I get it, surrounded by trees and set on the western shore of the Potomac, the CIA campus would be a cinch to ID from the air. The bad guys had hit our financial center, our Defense Department; driving a stake into the heart of our intelligence service would be the trifecta. It stood to reason that we had a bullseye on our roof.

But there was work to do, urgent work despite the growing footfalls moving down the hall outside the GRC door. It wasn't a stampede of fear, not entirely. For some, the fear was for families, the need to get home when the sky was literally falling. For others, though, it was just the reflex of obeying orders; get out meant get out now. But my career wasn't exactly a case study in following directions. With the evac order flashing in my face and cities on fire, it took little more than a heartbeat to know evacuating was not my preferred option.

Linda checked in with the team at the double-wide before returning to headquarters. On entering the parking lot, she found herself a salmon trying to swim upstream toward HQ, fighting against the flow of evacuees. Finally reaching the doors, she was confronted by a guard who refused to let her enter the building and ordered her to leave the compound. Somehow, she showed up in the GRC several minutes later, out of breath.

The room we occupied was the global comm center for the CTC. If you were to put a sign on our door it might read "in case of emergency, break glass." Separate from the Agency-wide Operations

Center that ran 24/7 rain or shine, the GRC spent a good part of its time in a mothball-like state of readiness. A small staff waited for something horrific, like the embassy bombings in East Africa or the attack on the USS *Cole*. When that happened, things in this room shifted into overdrive. We were designed to be the focal point of reach-back for anyone deployed in harm's way, people caught in a shitstorm—or out creating one. Whatever was going on today, it was a sure bet that things in here were about to get really busy. Hal and I rolled up our sleeves and started working the phones.

The media was reporting that the White House and State Department had also been evacuated. Our present reliance on cable news was unsettling. You can have HUMINT networks, satellites, and an NSA feed chock full of signals intelligence, but when a Godzilla-sized disaster was playing out on the streets of New York and Washington, DC, television was our most immediate source of information.

As catastrophic as events had been thus far, my gut told me that today was far from over. In a simple attack, a bomb or two goes off. At worst we have lingering threats from secondary explosives or armed assailants waiting to ambush responders. But the context of those factors are typically confined.

But this was . . . shit, this was a mess. Two towers with massive structural fires easily ten to twelve times higher than our ability to reach with hoses. I couldn't imagine an architect who would have thought to spec fire suppression for twenty thousand gallons of high-octane aviation fuel explosively delivered to the upper half of each tower. The entry and exit wound metaphor came back to mind; to some degree, those wounds cut completely through the building. That likely meant severing elevator shafts, stairwells, comms, and whatever fire suppression might be in place. At every pass, the media's best guess on possible casualties grew larger. Speculation climbed from a few thousand to tens of thousands.

The phone rang, the gruff voice of Rich Blee on the other end. He was with Cofer, bunkered in at the CIA's old print plant. Having been there recently, I had a clear mental image. It was a robust structure set slightly away from the center of campus, an ad hoc command center that was, it was hoped, just far enough off the X to make a difference if a plane dropped on our heads.

The prospect of that event gained strength. News came in from the FAA Command Center in Herndon, Virginia, just a short run up the road. A commercial flight out of New Jersey was reportedly in trouble. Air traffic control in Cleveland had caught screaming from the cockpit. Just before the plane went radio dark, somebody on board claimed a bomb was on the aircraft. The plane then wheeled around and was now on course back toward DC.

I looked back at the TV as the newscasters rambled on about the FAA grounding all flights, something I didn't recall ever happening before. The numbers were daunting. Out of some forty thousand aircraft in the sky at any given moment, how many of them were headed for something other than a designated runway?

Hal was great under stress, but Navy SEALs were trained to face the danger from the front row, not to be stuck in an office while family was alone in harm's way. The insidious nature of the attack turned my stomach. This wasn't a conflict between warriors but a matter of suicidal cowards attacking civilians. I couldn't begin to guess how many kids were on those planes. An awful thought hit me: did the WTC have a day-care center?

I picked up the phone, an outside line that didn't connect to the Pentagon or some command bunker. The call went home. A deep sigh slipped from my chest when my wife answered on the first ring.

"You okay?" she asked, knowing that the who and where of my work were rarely a matter of discussion.

"I'm fine," I replied. "Is Squirt with you?"

"She's right here." In spite of the stress her voice was level. Being home with a newborn meant she was thankfully outside of DC today. "Are you at the office? Are you safe?" We never referred to places by name; terms like *CIA* and *Pentagon* just didn't come up. But a fair bit of my career was spent at places that had either just been hit or, according to pundits, might be next.

She asked me what was going on when the television caught my attention, a wide shot of the two burning towers. For a split-second I thought what I saw was more smoke—gray-white as opposed to the oily black plumes that had been rising into the sky. This new stuff gushed out through the windows, cascading down the side of the building like a waterfall.

It took a couple seconds for my brain to process what my eyes were seeing—the top of the building was tumbling down as well. In what seemed like slow motion, one of the tallest buildings in the world just disintegrated. A tidal wave of dust and paper rushed down the streets of Manhattan, washing over a stampede of New Yorkers who ran in fear. Befuddled news anchors suggested that a piece of the building had peeled away. But as the camera looked on, only a single tower loomed up through the haze.

"I gotta go, honey. Stay indoors, and tell Squirt I love her." I spoke softly, paused, then added, "Tell Mike I won't make his wedding tomorrow. I'm not sure when I'll be home."

I swapped phones and punched the line to Blee. "You watching this?" Arguably that was a stupid question, but life thus far hadn't given me many chances to practice in the bizarro-world playing out right now. Rich came back with a profanity-laden expression of anger and frustration that could only come from a guy who had been screaming from the mountaintops that this day would come. When your nation is on fire, there is no comfort in being able to say, "I told you so." Rich was famous for saying it would take a thousand dead Americans before we got serious about bin

Laden. From what I could see on the TV, we'd already sailed past that number.

I briefed Rich on the missing plane out of Jersey. He came back with news from the Pentagon; on-ground reports confirmed it had been a commercial airliner. I asked if he had any word from Snake or Lovett. I could hear the weight in his voice when he said no. We all had friends in the Pentagon, and as the scale of the disaster revealed itself, it was likely we had all most likely lost some of them already.

Another line lit up, and I grabbed it. Mark was calling from the double-wide.

"93 is down," he said dejectedly.

I blinked, unsure of the reference until Mark elaborated. "UAL 93, the flight out of Newark."

"Looks like a pair of F-16s intercepted it as soon as it went AWOL."

The implication was horrific. "Oh Jesus, Mark, did we shoot it down?"

Mark chuffed, "No, not sure what happened, but 93 went erratic, then nosed down into the ground somewhere in Pennsylvania—as far as I know, in the middle of nowhere."

You know it is a shit day when losing an airliner in an empty field is not the worst possible news. I asked the obvious question: "Anything else loose up there?"

"Not that we know of. Everything with wings is being told to land at the closest runway. Word is most everybody is checking in and following ATC directions, but we'll be a couple hours at least before the sky is clear. What have you got over there?"

"Confidence is high that it was another commercial airliner at the Pentagon, but we're sketchy on casualties. I have nothing on Boyle, Snake, Lovett . . ." I drifted off, realizing the list of names was long as hell.

"Colonel Boyle's okay," Mark chipped in. "He was in Tucson."

The news was a much-needed lift, a tiny victory in the midst of a lot of shit.

"So who's left there with you?" Mark asked. "The parking lot looks like a fucking buffalo migration right now."

"Hal and I are here; Blee and Cofer have a team and are with the director in the print plant. At the moment, we're just trying to get a good handle on what's going on."

"Rog that. I have calls going out to the team; everybody is on order to double-time back here. I have logistics guys already loading up gear. You know the call is coming."

"Yeah," I said, a surly edge to my tone. "Only this time if we see him—"

"Fuckin' right." Mark growled, clearly sharing my mood. That was as close as we got to goodbye, and the phone clicked silent.

The STU-III line at our Air Force LNO's desk lit up. We punched the call on speaker and Snake opened without preamble. "What do you need, guys?"

Just as abruptly we explained that we didn't have the authority to task or obligate money, but we really needed the Predators on a plane headed to Afghanistan ASAP.

Snake didn't flinch. "Consider it done. What else?"

"We need Hellfires picked up from Huntsville and our team back on the East Coast to take the fight back to Afghanistan."

The call ended as tersely as it started, but I knew the message was received. Odds were that Snake had already made the calls to get a bird in the air while every other aircraft in the United States was being grounded. If anyone in the Air Force had the combination of authority and chutzpah to make shit happen in the midst of all this chaos, it was Snake.

A woman's voice caught my attention—not so much the frightened tone as the words she was saying. I turned to the TV.

"If you go over there, you can see the people jumping out the window. They're jumping out the window right now."

I watched the camera track a dark spot, arms and legs slowly cycling, as it plummeted down the side of the remaining tower. I tried to swallow and had nothing. In the midst of carnage on an incomprehensible scale, the human aspect is easy to lose. The options were hard to imagine—stuck in a tower a hundred stories in the air, no elevators, stairs ablaze, everything around you split between a furnace blaze and black-as-night smoke. Then the building that looks just like yours falls down, and you have front-row seats.

The visual gripped my spine. We train to fight, every bit of me wants to push back. But I wouldn't want to burn to death. A steady stream of falling bodies bore silent witness to those who chose to trade death by fire for one last, long, quiet step.

The collapse of the north tower didn't come as a surprise. With twin buildings suffering almost identical wounds, the outcome was inevitable. This time, cameras watched from every direction, shots coming in from the ground, other buildings, even helicopters. CNN had just pushed in from a wide shot when the upper portion of the North tower dropped into that slow, sick pile-driving rush to the ground, spewing an umbrella-shaped cloud of debris that fell with it. The massive column of smoke hanging above the tsunami of debris that rolled outward looked all too reminiscent of a mushroom cloud. I doubted that the aftermath of a small nuke could have felt much worse.

The shrill ringer of a satellite phone broke our focus on the second tower's collapse. Only one person had that number and it was my contact in Afghanistan. I answered the phone to hear condolences on the other end.

"We are hearing the news of an attack on New York City, and the Pentagon as well, my friend."

"We're under attack." I felt like Captain Obvious with that but had little else I could share.

"We are with you and standing by for guidance" was the response.

I told him it was going to take some time to get our plan together but that we were almost certainly going to be sending resources his direction and would need him as our partner on the front lines. With America all but certain to double down on our involvement in Afghanistan, the challenge of keeping the local resistance together just got a little bit easier. The odds that the remaining adversaries to the Taliban would be overrun were diminishing with each replay on TV.

Several minutes ticked by in silence. Not a phone rang, not a word passed between me and Hal. Footage streaming in from the WTC looked like Stalingrad—or the surface of the moon. One staggered, dust-caked reporter made a game try at continuing his report, distracted by a chorus of beeping noises all around him. Neither he, nor apparently any of the news anchors, had heard of a PASS device worn by emergency responders, a device that screamed for help when they were trapped and injured. My mind shouted, *They're around you; they're all around you*, as the camera swung across a pale gray backdrop of crushed cars and rubble. Some combination of cops, firefighters, and paramedics were buried, likely right on camera, hidden beneath the newfound landscape of ash and pulverized concrete.

I turned from the TV, muttered to Hal, "I'll be back," and headed out to the bathroom to flush the rising taste of vomit out of my mouth and run some water across my face. Halfway back I stopped in the center of the hallway, an inescapable oddity scratching at my unconscious mind.

I looked left, then right, apprehension rising until it hit me. It wasn't something, it was . . . nothing. On a floor where most rooms are referred to as vaults, an empty hall is no big deal. But it was alien for the building to feel empty, to feel dead—no muffled footsteps on carpet, no elevator dings, no thud of unseen doors or the hum of printers and copiers. As far as my every sense could tell me, the Central Intelligence Agency headquarters was a ghost town. We'd been run out of our house by an enemy who had yet to be formally named, but I'd bet money I knew what he looked like in infrared when seen from above.

A new sensation rose from my gut, a foul taste that I took for a moment to be fear or shock before I realized it was something completely different. I was pissed—sickened, frustrated, and damn angry.

13: PAYBACK'S A BITCH

MARK COOTER

AND THEY SHALL REAP THE WHIRLWIND
September 12, 2001

Nobody would mistake me for the bluebird of happiness, even on a good day. No one would sure as hell make that mistake when I'm exhausted, gut-sick, and desperately in need of caffeine. The Coke in my hand was working on that last point; the first two were beyond repair.

The notepad in front of me held what looked like a third-grader's homework. A laser-printed column of names floated in a sea of checkmarks, scratch-outs, and hand-scrawled notes. Much of yesterday and last night had been spent redirecting assets, locating the team, and talking to Ken Johns, who did much of the heavy lifting with the Air Force. I largely abandoned the usual check-and-balance part of how things were supposed to work. Anybody who even looked to throw a procedural challenge flag got a face-full of Unhappy Mark.

Though my directions were heartfelt and well-intended, it struck me in a brief lull that I might have gone to the well a little too often. At one point I told Alec, "I think I may have overstepped my bounds as a major in the United States Air Force." Whatever fight lay just ahead, I wouldn't be much help if I was in jail or lit-igation.

I made a call to Major General Shaffer to explain our plan. He found Lieutenant General Fogelsong, who ran operations for the Air Force. Shaffer came back on the phone and told me that Fogelsong was now aware of our plan and that whatever actions I needed to take were being done on his behalf.

I breathed a deep sigh of relief. Leaders sometimes wield great power, but great leaders empower the guys under their command to succeed. I was thankful as hell for the top cover and the trust it implied.

Darran had been at Langley Air Force Base glued to TV cover-age of what the press was already calling "9/11." In a brief phone call, he told me he would grab "Ski," an NCO weather specialist, along with any intel analysts he could lay hands on. They'd be ready to run as soon as I gave them a destination.

Colonel Boyle was on temporary duty in Tucson, Arizona. Absent flight options, he began his nonstop drive back to the East Coast with a little help from the highway patrol. All the while, he provided direction as needed.

DJ was out on a new job with a special-operations unit on Okinawa. I called and said, "I need you here now." I had no idea he was stuck in the middle of a typhoon. As a tribute to his determi-nation, he was on one of the first flights back to the States. Landing in Seattle, he found himself in a sea of chaos. Over the background din I could make out his shouted "terminal is packed . . . don't have a ticket." Up to my eyeballs at the moment, I had nothing to offer in terms of guidance. "Figure it out. Go to a truck stop if you have to. I need you. Just get here." That wasn't my high point in constructive leadership.

DJ took the straight-line approach, shouldering through a mob of pissed-off travelers to reach the ticket counter. He flashed his military ID and told the ticket agent he needed to get to DC to hunt down the son of a bitch who hit the towers. Moments later, he called back to say he was booked on the next flight. I made a mental note to hug him for his effort when I picked him up at the airport, whenever that flight took place.

Joker had called in from Alaska, his first words an anguished "Oh my God, we weren't quick enough." I told him to get to the East Coast as fast as he could, knowing damn well that like everybody else, he had to travel from Alaska to DC with no working air service. That's a tall order for most folks, but I had confidence a B-1 bomber pilot would find a way.

Genghis, one of my best pilots, had been on a weekend break when Gunny called her in the midst of the attack, opening with one of those "You watching the news? Any channel" calls. The two of them self-initiated on the spot, grabbing gear and letting me know they were carpooling to Palmdale. That seemed the smart play—if any military flights were moving from the West Coast, that's where they'd start. Not knowing at the time if they'd have a flight waiting or be forced to suffer some aberrant reboot of *Planes, Trains & Automobiles*, they set off at a sprint.

On the Agency side, Alec was keeping me up to date on the fast-paced decision-making inside CTC. He was in consultations regarding rapid decision-making on who should be on the first team into Afghanistan. The team would include two members from our program, Hal and Doc, who both had intimate knowledge of our capabilities. Most significant, Hank Crumpton was being brought back from overseas as quickly as possible to run the CIA's side of the war that was heading to Afghanistan. Alec was thrilled with the decision, as Hank was exactly the type of exceptional leader needed to take the fight to Afghanistan. Hank was smart, humble, and thoughtful in his approach to leading during his earlier tenure in CTC, and, importantly, Alec was confident we could tuck our program under his wing going forward.

I now needed to call the commander of the 11th Reconnaissance Squadron, my old boss. I told him I needed the pilots and sensor operators to get to Palmdale to meet an aircraft and they would be gone for an indeterminate length of time.

This was one of those moments I was directing action above my pay grade. His response was guarded. "Mark, I'm not going to get a formal message on this, am I?"

"No sir," I told him. "But General Fogelsong is aware, and I have his full approval."

That seemed to suffice. "Okay, if anyone else had called, I would have hung up. They'll be there. Good Luck."

As it turned out, Snake and Ken were able to work their magic in the flight department, conjuring clearance to fly a C-17 across the country in the midst of an FAA shutdown. I didn't want to ask how they pulled that one off, but I was thankful as hell to have them on overwatch.

With that in play we were able to pick up sensor operators Technical Sergeant Steve H., Staff Sergeant Andy R., Ken Mitchell, and Senior Airman Chris B., along with pilots Troy Johnson and Big. The moment the C-17 got a green light, everybody we had on the West Coast was redirected to Palmdale.

That was that, at least what was on my list. Everyone we needed was at least moving in this direction.

Thankfully, some of the key players were already here. For them the nightmare of full-rush travel was swapped for the abrupt need to juggle flaming chainsaws five at a time. Cliffy was at the bleeding edge of that stampede, working like a man possessed to finalize or dress-down the wiring in the GCS and the double-wide.

I looked at the list in my hand. The words on the page blurred, and I rubbed my eyes. Two days ago, we felt like we were leaning forward, making progress ahead of authorizations and budgets. Then 9/11 hit, and everything had to be full-on, not "sometime soon" but "right fucking now." Nothing was ready for prime time, but Cliffy was on a no-sleep cola-fueled marathon, shoulder to shoulder with Paul Welch, to close the gap.

Ginger was in the thick of it, collecting everything she could lay her hands on about the posture of the Taliban military, what al-Qaeda in Afghanistan was doing, and what the US military was doing. Many of our new crews had never operated in Afghanistan or even in a highly contested environment. Our data from 2000 was already outdated. All our procedures and checklists needed to be brought up to current. Charging forward under the banner of "right fucking now," Ginger became an intel juggernaut, assembling intel and operations information so we'd be ready to fly.

The C-17 from Palmdale touched down at Andrews Air Force Base and began unloading people, cargo, and Spoon's rental car, which had come along for the ride. Darran immediately started running sheepdog, wrangling a blue Air Force school bus to shuttle the team to the hotel.

I took a long slug of Coke and stared at the checklist. Despite the frustration and lack of sleep, I couldn't be any prouder of the team. Every one of them was all-in, dropping everything to get back in the fight. I had no idea what personal sacrifices were made, only that they were numerous and, in some cases, likely costly. In the greatest tradition of American patriots, they all ran toward the sound of gunfire.

My eyes drifted further down the list where inbound people gave way to inbound things. Prominent on that list was the impending delivery of a case of jalapeños, open-air code-speak for Hellfire missiles from the US Army inventory. We were getting a limited number of Kilos now, with a mix of more Kilos and Mikes coming later, reflecting the older K and newer M models.

The Ks were designed for the missile's original role as a tank-killer. The tandem-charge high-explosive antitank round could burn a hole through plate steel armor. But as we'd seen in Taco Bell, firing what amounted to a laser beam into a room may punch a smoking hole in the ground without having as much effect on the surrounding area. Without an actual hit, the odds of surviving a Kilo out in the open were comparatively high.

The Mikes were a different animal altogether, created when the role of Hellfire was expanded to include ships and small bunkers. The Kilo's HEAT warhead was swapped out for a blast/frag/incendiary charge that lived up to the name Hellfire.

While the Air Force guy in me had a soft spot for dropping two-thousand-pound JDAMs a pair at a time, there was an argument for a weapon you could shoot through a window that didn't level a city block. Although the Navy had seen only mixed success using Mikes against small boats, we hoped they would do better on our potential targets.

POKING THE BEAR
September 18, 2001

One week to the day after the towers fell, Predator 3034 sat on a familiar Asian runway. At 2330 Zulu (or Greenwich Mean Time), the adrenaline-fueled surge of the past seven days came down to the nose camera image as Predator rumbled down the runway and rose into the sky.

Either in response to current intel or maybe a bit of wishful thinking, the first mission was a lap back to Tarnak Farm where we'd last seen Usama bin Laden. As improbable as the odds might have been to find him there twice, nothing more was on the forefront of our minds than to get one more glimpse at that tall blob of infrared—well, that and to shove a Hellfire down his throat.

Sadly, nothing in war is ever that simple, and while most things looked the same as it did when we last visited, UBL was conspicuously absent. We executed the remainder of the mission in relative quiet, gathering photos and video of any changes that may have taken place during our absence.

October 6 was to be our final shakedown mission before Operation Enduring Freedom was to kick off. For this mission, I instructed Darran that he had a secondary mission to the one that Alec would give us. I told him it was our time to poke the bear and maybe tickle it a little.

The bear in this case was neither my temper nor an Afghan grizzly but another damned SA-3. If there was any upside, we had pretty good maps of the radar installations and had an even better idea of the radar coverage. We knew where terrain obstructed line of sight and where radar could see best.

I came into work rested and ready to prosecute our part in the invasion. I opened the door to the double-wide and could see lots of activity like a team getting ready for a big game.

I could also tell something dramatic had happened while I was away. My eyes met Ginger's, and she said, "Oh my God." Before I could blink, Darran came up with a wild glint in his eyes. It was all I could do to blurt out, "What?"

He replied, "They launched an SA-3 at us. Don't ever do that to me again."

A sudden foreboding gripped my heart. Had we flown one of our limited Hellfire birds? *Whew!* No, I remembered this was a so-called classic bird: no MTS, no Hellfire. I was the one who sent us into the bear's den. My next words came out remarkably calm given the adrenaline hitting my veins. "Did we survive?"

Darran blinked, realizing at once that, unlike everyone else in the trailer, I had no idea about the end of the story.

"Oh, hell yes!" he said proudly, eyes now truly ablaze. As only Darran could, he added, "It was cool."

Swallowing what had momentarily threatened to be a second taste of my breakfast, I called for everybody who had been present to debrief me on the incident. Half a dozen bodies crammed into the room, eyes alight with excitement.

I took a shot a kicking off the conversation. "OK, so when I left last night we were on a nice, quiet search for some Christians being held hostage. How the hell did that evolve into having a missile shot at us?"

The assembly all turned to Darran, an animated storyteller, who described Joker's closing in on the SA-3 radar perimeter like Steve Irwin described creeping up on a croc with a stick. We knew the SA-3 was out there; it was one of the things we kept tabs on whenever we were in the neighborhood.

Darran explained that we had come in from the southeast, keeping as much distance as our sensors would allow. Andy had been on SO, scanning the area for any sign of missile crews or, worse yet, a missile launch.

"We got nothing," Darran said, "so I told Joker to tickle it like you said. That was when things got, hmm . . . interesting."

Interesting is never a good word in air-combat missions. I knew that Chris J. would have been monitoring our threat display and chat channels. His job was to watch the bad guys and listen to bad-guy chatter in case they saw us.

Darran continued, "We started carving successively deeper 'scoops' into the invisible dome of radar coverage. That's when word abruptly flashed from the double-wide that we'd been made."

I could imagine that moment in my mind. Joker would have had only a few choices. Break and haul ass might have been high on the list, but a race between Predator and an SA-3 would have been on par with a shopping cart trying to outrun a Corvette. Flop down on the deck, try to get lost against ground clutter . . . I ticked through several alternatives.

"So what'd he do?" I asked, seeing that my crew of air warriors were clearly enjoying their side of the delivery. Something told me this was going to be good.

"He played chicken." Darran said, snapping his fingers in the air. After sharing a conspiratorial grin with the faces around the table, he decided he'd milked the dramatic tension long enough and let me peek behind the curtain.

I nodded, knowing he was probably right. My impatience was rising. "And?"

Darran said, "That's when the Low Blow went active."

That drew a scowl, despite the fact that I knew the story ended well. Missile sites use a combination of radars to find, fix, and finish off a target. The Low Blow is the fire control radar that guides an SA-3 to its target. A Spoon Rest radar is the one that has the best chance to detect you in a big open sky, but the Low Blow is the one that'll get you killed.

Darran continued, his level of excitement rising. "MiG had everyone watching the live feed of the radar to see if it started hunting. That's when Chris shouts from the double-wide claiming his sensors picked up a spike in radar emissions. Somebody called bullshit, but Marcella[65] and the imagery team pulled up video and sure as shit he was right."

MiG cuts in on the narrative. "So I had them haul ass straight for the radar from a different direction."

---

65 Staff Sergeant Marcella Avans was our lead imagery analyst. No matter the situation, she always had a smile on her face. One of the newer members of the team, she quickly became a trusted junior leader. Marcella provided real-time backup to our sensor operators, helping them identify what we were seeing in various spectrums of light.

As bizarre as that might sound, it was the right call. In a lot of cases, radar coverage is more like a doughnut than a dome; it has a max range but also a minimum one. If we could get inside the doughnut hole, we could disappear from the radar screen. The vulnerability varies from system to system, and MiG knew exactly what to do to exploit this one.

I nodded. "Did it work?" I found myself getting sucked into the story.

"Hell yeah, we min-ranged it," Darran said, pointing at me. "MiG is the man. We parked in the blind spot right over the dish. Nobody is shooting at us, but we're stuck inside the doughnut hole. On the plus side, we have a tank full of gas and time to consider our options."

Time was a rare luxury in air combat scenarios. Jets suck fuel out of tanks in great gulps, hurtling past a target at speeds so fast that details blur. Predator, on the other hand, put-putted around in a slow circle, sipping from the gas tank more like a John Deer lawn mower than an F-15.

"We decided to call in a lifeline," Darran continued, "so Rich pings the guys at MSIC[66] in Huntsville and asks for their top SA-3 brains. That was sorta funny—they wanted to know where we were and why we needed the data we were asking about. We came back with, 'Never mind, and answer the question,' which seemed to get us by. They told us that if anybody tries to crank a Low Blow far enough back to look straight up, it'll go top-heavy and actually fall over. The only way they can make it work is to have a team of technicians manually unlock some retaining pin." Darran grinned,

---

66 Missile and Space Intelligence Center. MSIC is part of the Defense Intelligence Agency and has among its missions to exploit foreign missile-related technologies to find vulnerabilities and countermeasures.

"We were seeing a bunch of foot traffic milling around the radar, and smart money said they were yanking that pin."

"Awesome!" I had to admit, that was good. "And then . . ." I prompted.

"Well, the Afghans were going batshit, and we had the best seats in the house. Joker kept us in this tight little orbit while Andy scoped out everything. Facilities, vehicles, launchers—hell, we collected so much of their emergency-response activity it probably makes for a pattern-of-life report. Hell, there were TV reports on them shooting triple-A[67] at us."

It was Ginger's turn to pick up the story. "So, in the middle of all this shit the phone goes off, and it turns out to be Lieutenant General Gene Renuart, the USCENTCOM J3. He wanted to know what the hell was going on, and, honestly, I just didn't have the bandwidth. I told him, 'Sir, unless you know of a way to defeat an SA-3, I gotta go,' and hung up."

I tried to bite back the grin. I'd known Renuart maybe fourteen years; he'd been my commanding officer twice. I could just imagine his face when some captain hung up on him. This was getting better by the moment.

Silent thus far through the story, Joker chimed in. "What I was worried about was TV coverage of getting shot out of the sky. Since MiG figured us a way into the doughnut hole, I told him it was time to figure a way out."

Darran looked at Joker, then seemed to settle just a bit, his serious side resurfacing. "Of course, we couldn't just fly a tight circle forever; our gas would run out at some point." Then his face brightened. "That's when it got really cool! It was Troy's idea."

---

67 Anti-Aircraft Artillery

That piqued my interest. I thought pulling a vanishing act in an empty sky was a damn good trick to begin with, so I turned to Troy. "This I want to hear."

A soft-spoken guy, Troy briefly returned the smile before becoming thoughtful about his response. A pilot from General Atomics, Troy had climbed through pilot ranks to become instrument rated, instructor rated . . . his list of accomplishments went on. That took a special kind of mind, someone with amazing attention to details, numbers, and procedures.

"It was sort of an escape-velocity thing," he began.

The term caught me by surprise. Did Troy have NASA creds I didn't know about?

Before I could ask, he filled in the blanks. "It was sort of an Apollo maneuver, the one where they slingshot around the moon. We had this small circle to play in, and I suggested that Joker punch the gas and start building up as much speed as we could so we hit our escape point at a dead sprint. Even in that tight bank, Andy held the sensor centered on the radar and launchers."

I shook my head in awe of the out-of-the-box approach, one you'd never find in a flight-ops manual. I also made a mental note that we'd add it to ours.

Darran jumped in, "MiG passed us the best egress heading that the big brains in the double-wide could come up with. So, I said, 'OK, let's hike our skirts and get the hell out of here.'"

"Joker was great," Troy continued. "We hit something like 180-plus in terms of ground speed by the time he broke for the hills with the wind at our back. I'm sure we lit up the screens like a Christmas tree at that point, but by the time they could pull the trigger we'd put over twelve miles between us and the launcher. Andy held the ball steady and had eyes glued on the site the whole time so we could watch the missile come off the rails. The first one failed on the rail; that was a huge break. The second one must have had some sort of launcher trouble, or got distracted by ground clutter, because it left the tee, took a wicked slice and sailed off about 135 degrees off-axis."

An officer in my shoes might have felt a lot of things at that moment, relief likely high on the list. For starters, it had been my call to poke the bear, and had that gone badly, the blowback would have landed squarely in my lap. But watching this circle of men and women, each pointing out what their teammate did right, how the person to their left or right made all the difference, I felt only pride. Out in the "real" Air Force, life is governed by things like rank and role. Maybe it has to be for something that large to work. But here in this cramped little outpost, what Alec called the Island of Misfit Toys, we had a team . . . a shit-hot team. We may have had different temperaments, talents, and convictions, but we had a common bond. Shared ideas and shared credit—everyone's eyes were fixed on a common goal. And I'd never before been so happy to learn that a missile had been shot at us.

Of course, information isn't much use if you don't share it, and I knew some guys to whom this whole experience would be particularly valuable. I made a call to the B-2 unit at Whiteman Air Force Base in Missouri. I knew they'd be the first guys flying into Afghanistan and would be way behind the power curve when it came to the Taliban air and air-defense capability. Unlike an unmanned flight, the B-2 flight crews had their ass on the line if a missile came running, and they deserved every edge we could give them.

As our earlier brush with USCENTCOM had taught us, I didn't have much confidence that they would move spritely to pass along the details. That means skipping the front door. Luckily, the wing's chief of intelligence, Major Mike "Rabbi" Harasimowicz, was an old friend and had worked for me back when he was a second lieutenant.

"Rabbi, it's Cooter."

"What's up?" he replied.

I dispensed with the pleasantries; some could lead to questions I didn't want to answer. "I'm sure you've got a lot of questions about Afghanistan, and I think I can answer some of them. I'm going to put some of my guys on the phone. You might wanna have a notepad."

I handed off the phone, and over the next several minutes my guys gave Rabbi a full-throttle data dump on Afghan radars and their operating procedures. When they finished I took the phone back.

"Useful?"

He replied, "Holy shit, that was exactly what I needed. How do you know—"

I interrupted him and said, "Sorry, gotta go. If you need any-thing else . . ." I gave him my classified email address, said, "Good luck," and hung up. That would answer some of Rabbi's questions and likely raise a dozen others.

A big sigh rolled out as I turned back to my desk and opened the attack plans for a strike on one Mullah Omar Mohammed. It now was October 6. Though the world didn't yet know it, the start of Operation Enduring Freedom was just hours away.

AFTER OMAR
October 2001

While the strike on Omar's detail didn't cut the head off that par-ticular snake, it proved beyond any doubt that Predator could be pinpoint lethal in a no-shit combat environment. There is an old saying that failure is an orphan, but success has a thousand fathers. After just one engagement our ugly little child started seeing adop-tion papers.

Like so many things thus far, this was a double-edged sword. Though the unfamiliar surge of interest from within the top-secret community was rewarding, our infrastructure was still held together by Velcro and zip ties. We were inventing fixes for our own problems on a day-to-day basis, receiving software updates almost nightly. That was a long-ass way from being able to prosecute some other mission on somebody else's timetable.

Colonel Boyle continued to be the iron curtain that separated us from our detractors, the naysayers, the people who for reasons earnest or political wanted to see us shut down or fail. That group included politicians and general officers, a level of military society that colonels with career awareness are encouraged not to piss off.

But Colonel Boyle never flinched. When told in forums both private and public that we couldn't possibly succeed, his answer was not just that we *could* but also that we *are*. Some days, it looked to me like the professional equivalent of doubling down on a pair of sevens, but he believed in us, believed in the mission. Time and again he played "you bet your ass" on the fact that the Air Combat Command Expeditionary Air Intelligence Squadron would be the first Air Force unit to fire a Hellfire in combat. Now we had added that to the growing line of notches on our stick.

Although we were here to put eyes on bin Laden, within a short time we had both Agency and military people on the ground and in need of support. We recognized Predator as an amazing tool to support these elements but were really unable to communicate with them or share time-sensitive video with anyone but those in a few key command centers. We could see it half a world away, but the US people on the ground couldn't benefit from video being captured directly over their heads.

Who better to tackle this problem than the same geniuses that gave us Predator, RSO, and the Hellfire capability in the first place? Once again it was Big Safari to the rescue. The requirement that went to Bill Grimes, Brian, Spoon, and the rest of the team became to give us a capability for elements on the ground or other aircraft to see what Predator was looking at. The result was an understandably clunky antenna, video monitor, and power supply to intercept the video being transmitted by the Predator. I would have gone with "magic."

Once at altitude, the aircraft was flown by satellite via the Ku-band link beyond line of sight of the LRE. Since the C-band link wasn't used downrange, the pilot would now dedicate the C-band for the ground element or other aircraft to view the video, giving them situational awareness and the ability to share a common operational picture of the battlefield.

The tool became known as "Rover," and it gave others the ability to have a God's-eye view of the battlefield. This was a substantial evolution in the value of Predator.

Even greater was the ability to place a Rover terminal in aircraft like the AC-130. Now, instead of our pilots in Virginia talking the gunners of aircraft onto the right target, the gunners could have their own monitor and more precisely receive target designations without lengthy talk-ons, avoiding possibly disastrous misinterpretations of directions. The enemy perceived it to be a very bad thing when what they termed the "bumblebee" and the "buffalo" were in the sky together.[68]

But if there is one constant in the world of aerospace, it is that technology hobgoblins are a persistent threat. Old problems return, and when we finally fix one a new problem pops up in its place. A notable goblin reared its little green head while we were in the midst of lazing a target. Firing a beam to lead a missile to its

---

68 In both HUMINT and SIGINT reporting, the Taliban and al-Qaeda referred to the Predator as the "butterfly" or "bumblebee" and called the AC-130 the "buffalo."

target, the laser unexpectedly shut down. That's *not good*. Without the laser, we had no fangs.

The problem was attributed to a power overdraw, the laser suddenly sucking a few more amps than the electrical system had to spare. In an aircraft, electrical power is something of a zero-sum game—if you exceed your limit by so much as an erg, something shuts off.

The great history of NASA pointed us towards a corrective approach. During the crisis management of the Apollo 13 rescue effort, astronaut Ken Mattingly led a ground team that faced a similar hard-stop limit on electrical power. With the weight of three lives on their shoulders, the ground workers repeatedly went step-by-step through the power cycles, carving needless draws from the system. When they exhausted the needless they turned their eyes on "the important things we can survive without for a few minutes."

Taking up that banner was Big and Phil M., the latter the smartest tech I knew from General Atomics. They led the General Atomics tech team, supported by Raytheon and L-3 engineers on-site and back at their respective plants, and started plowing through the manuals and procedures. When an aircraft has no human occupant, it automatically saves the enormous costs of power, space, and weight needed to encase a human in an armored shell and provide an atmosphere. But that simplification also left considerably fewer systems from which to select a sacrificial lamb. Most everything on a UAV was essential to its flying.

We turned off power to lights and momentarily turned heat off to the pitot tube. In a brilliant flash of insight, Steve H. hit on an answer that had escaped everybody else—the parking brake. Absent a pilot to pull the parking brake lever, Predator's parking brake was kept in the off position during flight by a small electromagnet. That magnet drew electrical power.

Steve H. delivered the idea with a simple analysis: "It's not like we care if the parking brake is on when the wheels are off the ground." Flight procedures were adjusted, and those precious amps were redirected to feed a hungry laser. The beam came back online.

THE BEST DAY
October 21, 2001

"You're kidding me. The same house we spent eight hours staring at yesterday?" I closed my eyes, consciously dialing back the frustration.

If Alec noted the irk in my tone, he shrugged it off. He'd been around me long enough to know that if I was really bent, it would be evident. He replied calmly, "We're working a USCENTCOM requirement, and the house outside Kandahar keeps coming up."

"Roger that," I said, chiding myself for the momentary lapse. Primarily, we were a CIA surveillance platform, not an Air Force gunship. Yes, we now had weapons, but that didn't magically make Predator some sort of big-deal close air-support platform. We had a boatload of those: F-16s, F-15Es, F-18s, and B-1s. By comparison, we were just a spy plane with a flyswatter. Still, I grumbled to myself, *I'd be really happy to come across a couple flies out in the open.*

Setting aside the déjà vu, replicating yesterday's flight to Kandahar was otherwise reasonably quiet. With repetition, we had reinforced our operational knowledge of the flight corridors, gaining from experience the comfort that we could disappear into the rugged terrain the way a stage magician steps behind a curtain.

As the flight clock slid slowly around the dial, I pondered the reasons for the second lap. Alec mentioned requirements from USCENTCOM, but that was a catch-all term for the small circle of authorities that knew we existed. Charlie Allen and Tommy

Franks sat at the top of a very short list of people authorized to pick up the phone and say, "Can you look at this?" Those requirements are pooled, then sifted for a good match with our platform. Despite who is asking, some requests will trump others. US guys in harm's way went to the front of the line. If an American was getting shot at below our piece of the sky, we'd come at a run.

The stuff asked of us, the things we were best suited to do, were prioritized into a collections deck. The deck established our mission parameters for the day: take off at such a time, go to this location and conduct pattern-of-life collection for some duration, and so forth. Some missions had several short milestones; others became an extended park-'n-watch to map out when buildings were occupied or empty and what cars parked around them, all clues to daily activity. If we wanted to know when something was up, we had to establish what normal looked like.

We flew every moment we could—a Predator on the ground is an expensive camera tripod. If gaps in a mission came up for one reason or another, we'd lap back on a prior site or swing wide en route to the next, snatching glimpses of secondary targets. If something hot came up unexpectedly, the deck was shelved, and we'd have to wing it. That typically began with a call from Alec or Hal relayed by DJ, Eric, or Ty, depending on who was on shift.

I glanced at the clock: 0243 hours. We'd been circling over the suspected Taliban safehouse in Kandahar when the phone rang. It was DJ. I could tell from the sound that he was typing and writing in parallel, working his own notes while pushing data into the team-wide chat system. I leaned to see my own console and saw two words in a growing block of on-screen data: "Tarin Kowt."

"What's up?" I asked.

"SIGINT's got something." More clatter of keys followed as DJ multitasked. "NSA scored a hit on a known device. They're running voice analysis now. Alec's about to ask for transit time to this location." He thumped what I took to be the Enter key and pushed me a set of grids.[69]

"Alec's gonna want . . ." I chuffed at DJ's predictable flash of prescience. Like Radar O'Reilly from *M*A*S*H*, a great LNO is always a step ahead. Before I can finish asking for a form, he hands it to me, already filled out. DJ knew every gear in the intricate clockwork of two competing cultures, Agency and Air Force, and bridged them seamlessly. If I had to guess, I'd put Eric or Ty, maybe both, running support on their own initiative. For three guys who were supposed to work in shifts, it was more common than not to find all of them working at any given moment.

I looked at the screen, mentally running the tables for the distance to Tarin Kowt. Having a pretty good sense of the map, I tapped "swag 45–50 min" into chat and pushed it back to DJ. That wasn't the scientific answer, but it would put him well within the ballpark.

After a brief exchange, DJ came back on the line. "OK, it's official. We have a credible SIGINT cut. Alec wants us to break contact here and go take a look."

Before I could ask my next question, DJ added, "They'll have a refined ellipse by the time we get there."

"Do they have reference imagery?"

DJ came back instantly. "Do they ever?"

I swallowed the grumble that rose in my throat. "I'll go check with Marcella." The grids were already team-wide on the chat system, so everyone in the GCS and double-wide would be working their slice of the pie. The team would be pulling up FalconView

---

69   The term *grid* refers to the way the military defines and communicates a geographic location anywhere on the planet. In practice, it is like laying graph paper over a map and defining a point as the intersection of X and Y values.

data along with any satellite or aerial-surveillance photos tagged to those coordinates.

Under normal circumstances, getting a tear-line report from the NSA citing activity at a given grid was enough to justify a look, but when we intend to deliver ordnance on a target, the details become critical. We were on the back end of a steep learning curve at this point and knew to ask the right questions to determine the time and accuracy of the grid.

In this case, the collected information contained an inherent error probability—in layman's terms, an oval so long by so wide, within which the signal originated. The larger the error probability, the wider the area of uncertainty. In a perfect world, we'd want a circle one foot across; in this case, we were looking at an area much greater than that. Intel is never perfect.

We would need to wait until we were over the area to find out if we were dealing with a rural or urban environment. Rural was a challenge; absent some fluke good luck of a vehicle visibly parked in a grass field, rural settings offered very little in terms of defining a target location with specificity. Terrain features were often subjective—imagine trying to talk a pilot in on a clump of trees in rolling hills smattered with clumps of trees. Double that at night under infrared imaging.

Heavy urban was the opposite challenge, both having to identify one building among a carpet of others and dealing with the risk of collateral damage. There's no foul if you hit the wrong tree in the woods, but hitting the wrong building in a town could be horrific, or made to look so. An abandoned warehouse will, with little doubt, have a hand-scrawled sign that reads ORPHANAGE thrown on the rubble by morning.

A suburban setting was as close to ideal as one could hope for: small clusters of distinctive structures separated from one another to the point that they could be visual reference while remaining outside of frag radius. The countryside around Tarin Kowt had potential; it was largely agrarian with small, walled-in compounds separated by fields of crops.

I nodded to nobody in particular and ran through my own checklist. The pilot and double-wide would work together to get air clearances and asset availability thru the CAOC[70] or some other airborne asset. That'd take ten to fifteen minutes. Ginger had the rest of the team to get weather, target info, and known threats.

I turned to the pilot and sensor-operator stations. Joker was flying, with Leo on sensor. Big was working as second pilot—a fancy term for being stuck in a cheap folding chair wedged between the pilot and sensor-operator seats.

As I expected, Joker was working the time to the new grids. He and Big went back and forth in a verbal version of pilot shorthand, a concise exchange of acronyms and data that, while gobbledygook to the layman, spoke volumes to the trained ear. Big glanced back at me and waved to get my attention.

"We can run a pretty direct path; we have the airspace. Winds are not helpful." He shrugged. "Could be worse. All in all, we're fifty-five, maybe sixty minutes out. Depending on how long it takes to make a decision, we should have plenty of fuel."

I nodded, running the math through my mind. Every minute of flight time had a value, which meant burning a minute had a cost. This little treasure hunt would chew up four hours we could have spent gathering data on a known location. "DJ, it looks like we're about an hour out from the target."

---

70 Combined Air Operations Center. Among many functions, it serves as the traffic cop for air combat.

In the spirit of "nothing travels faster than bad news," I barely had a chance to breathe before DJ hit me back, leading off with a chafing, "Alec asked if you can pedal that thing any faster."

"Hah, hah." I wasn't a big fan of ad hoc detours, often as not an expensive wild-goose chase. I fired back, "What's up anyway? You guys spin the Wheel of Random Locations again?"

Alec picked up on the line, perhaps to shrug off the counter-punch in person. "We're working an intercept just a little north of Tarin Kowt. From the sound of things, a large group of al-Qaeda are rolling in for a face-to-face. NSA got the hit off our bad-guy list of phones, and Roger is working it now to confirm from the voice cut that it is actually our bad guy on the phone. He said security personnel on the ground are being directed to square away parking for several vehicles, set up a perimeter."

I nodded, processing the data. This had the hint of promise. "Big group. Sounds like it could be somebody important."

Alec grumbled under his breath, then added, "Let's hope for really fucking important. Hang on a sec . . ." His voice abruptly turned away from the phone. I heard the tones of a short exchange and Alec returned. "Okay, Roger has a 90 percent confidence on who made the call. He's on our known bad-guy list. He's on a second call now, and his signal doesn't appear to be moving. If this pans out, he should have company pretty soon."

"Good deal," I said, my brain shifting to the immediate tasks. "We're already en route, looking at what other assets may be available if needed. If it's gonna be a party, we'll wanna bring friends."

"Copy that," Alec replied. "I'll push intel as it comes in." The phone clicked silent.

I turned, then pinged DJ. His voice came over the speaker, in a rapid exchange with what sounded like Eric. "Data coming your way," he fired off, the sudden clarity suggesting he was speaking to me. I looked at my monitor and scanned the list that appeared in chat. Two F-15 Strike Eagles were working close air support for an ODA west of Kandahar.[71] The Navy was running F/A-18s off a carrier in the Arabian Sea. Compared to the numerous trade-offs that led to Predator itself, the options we had to draw on could be numerous on a good day. Push-CAS is a beautiful thing.

Most people think of an air-combat mission in terms of planes flying from here to there with a specific goal or scrambling into the sky à la *Top Gun* to tangle with enemy fighters that come to visit. That's the Hollywood version.

The real core of US airpower is the persistent advantage of high ground. As a rule, we maintain an armada of air assets, fighters, tankers, command and control aircraft, and ISR aircraft overhead to serve the immediate needs of the warfighter. Soldiers under fire rarely have the luxury of waiting two hours for air support to arrive. It can be a game-changer when a hulking AC-130 gunship, bristling with weapons, comes out of the clouds to rain laser-like streams of hot lead on an enemy. But you have to own the sky to make that happen.

Push-CAS is a strategy that saw a renaissance during Desert Storm, but its roots traced back to World War II. The idea is to push a relentless stream of aircraft into the battle space, day and night. For every plane landing, another is taking off. For every fighter turning back to home, another is hurtling into the fray. Of

---

71 ODA, or SFODA, refers to a Special Forces Operational Detatchment Alpha, or A-Team, the primary fighting force of the US Army Green Berets. Typically a dozen highly-trained men who operate in the wild of a war zone with minimal external support.

all the ways to dominate a conflict, from advanced technology to superior training, a lot can be said for overwhelming numbers.

I stood, arms folded, watching the thermal-image screen as moon-gray terrain silently slid by. The ground was a patchwork of sown fields, irregular pieces of quilting stitched together by a dotted row of trees or a meandering dirt road. Every few hundred yards, we saw the typical Afghan compound, typically a small home with a couple out-buildings enclosed in a high medieval wall. Out here, every man's home really was his castle, at least as close as you could get with mud and sticks. Between the Taliban, al-Qaeda, and the rollover of local warlords and bandits, the people only get to keep what they can defend.

We rolled up on the coordinates that defined the ellipse. The camera stared down through thousands of feet of sky at a compound that, in most aspects, looked like every other one. Decidedly atypical, though, were the dozen vehicles that surrounded the central structure, a decent-sized meeting house. Dozens of blobs moved around the compound like ants, some alone, most in groups. I didn't need to be an image analyst to see the pattern of guards walking a hasty perimeter.

Joker settled the plane into a smooth orbit, giving Leo a stable platform from which to scour the area. I watched the movement on the screen, knowing Alec was staring at the same live feed. But tucked off in the GRC, Alec couldn't listen to the exchange between pilot and sensor op, the pros with their hands on the sticks. That, as radio's Paul Harvey used to say, was the rest of the story.

"I've got lots of personnel." Leo had pushed the zoom on the Raytheon ball. In white-hot mode, living flesh glowed bright against the gray dirt.

The people milled about, some holding post at a gate or around a vehicle, while others plodded haphazard laps. Occasionally one blob would move rapidly, arms motioning in one direction or another, setting a meandering group on a new direction.

I shook my head. *If your being in charge is obvious from twenty thousand feet up, you're doing something wrong.* The energetic blob, a leader among lesser blobs, promoted itself higher up on the target table.

The running analysis blurred into the background as I shifted my gaze to a different screen of black-and-white daytime photos flagged with notations that linked it to AQ operatives. A couple had been pulled out of FalconView, with some retrieved from the GRC, along with other sources. Marcella had drawn connections between the various images, identifying consistent features. By combining the different views and spectrums, we knew more than any single one could provide.

The clock ticked, and a slow twist began to wind its way through my gut. The Agency guys would apparently consume every available second before actually making a decision. As a rule, that left no time at all to actually act. But finally, we got clearance.

When the green light hit, I looked to see who was stepping into the batter's box. I grinned when I saw Eagle One-Niner, an F-15E Strike Eagle loaded with GBU-10s (Guided Bomb Unit-10s, laser-guided bombs). Perfect.

The GBU-10 was the big dog of the Paveway II series. At its heart, it was little more than an old-fashioned Mk 84, a two-thousand-pound "dumb bomb," the likes of which we rained across Europe during World War II. What brought these Mk 84 into the twenty-first century was a high-tech nose cap that added a complex package of laser-seeking and control fins. This transformed the weapon from one that landed "wherever the hell it fell" to one that could be laser-guided into a three-foot circle.

This remarkable accuracy had been demonstrated in 1991, in a wildly unconventional fashion, during close air support of a

Special Forces team during Operation Desert Storm. Diving from the clouds on a group of five Soviet-built MI-24 Hind gunship helicopters, an Air Force F-15E engaged not a ground target, but a helicopter with a GBU-10. Although the helicopter was already moving at some one hundred knots and climbing, the F-15E guided the GBU to a clean hit that, understandably, vaporized the Hind. That was an extreme instance, but by comparison, our target tonight was the broad side of the proverbial barn.

"Eagle One-Niner, this is Wildfire Two-One, we have a target at grid . . ." Joker rattled off the grid location, followed by a concise verbal description of the compound. He wrapped with a declaration that our laser was dancing a bright spot on the roof.

My internal countdown to impact was already running when the Eagle pilot replied. "Negative, Wildfire, we don't have your target."

My head snapped around, jaw slack. *Say what?*

Joker flashed me a raised eyebrow but continued on the mic. "Uh, what part of that don't you have, Eagle One-Niner?"

The response was definitive. "None of it, Wildfire. We're not picking up a laser, and from where we sit, every compound in the valley looks the same."

Not picking up the la— "Aww fuck," I growled. "It's the LANTIRN." Around the GCS, heads swiveled to face me. I clarified: "It's the fucking LANTIRN."

Despite my eloquence, only a couple of the faces showed the click of connecting on a matter of arguably obscure Air Force technobabble. The AN/AAQ-14 targeting pod, aka LANTIRN[72] couldn't see our laser marker.

---

72 Low Altitude Navigation and Targeting Infrared for Night, a combined navigation and targeting pod system originally developed by Martin Marietta, used on the USAF's premier fighter aircraft like the F-15E Strike Eagle and F-16 Fighting Falcon.

Joker switched gears in a heartbeat, abandoning the laser in favor of an old-school "talk on."

"Eagle One-Niner, from your heading there is a prominent dirt road running east to west, taking a mild northwest slope before flattening out due west again. Your target is the second compound in the northwest angled stretch of road, located on the southwest side just after a long stretch of dark wheat field."

I listened to the exchange, realizing just how differently we viewed the same battle space. Our view had more in common with that of a helicopter. While not parked in the sky, we were certainly driving slow circles in first gear. If the Eagle overhead slowed down to our top speed, it would stall and fall out of the sky.

In contrast, Eagle One-Niner's view of the compound was more like a Formula 1 driver's view of pit row: "I have a gray blur smearing into darker grey streaks."

Fifteen minutes later, Joker had described the scene to the point that an artist could paint it. Whatever view Eagle One-Niner had of the terrain, no number of adjectives was going to close the gap before he ran out of fuel. With no laser and no talk-on, it was time for Plan C. For that, I needed to talk to Alec.

He picked up the phone with "I was expecting something with more kaboom."

"Yeah, yeah," I said, shoving past the obvious, "We've got a disconnect on delivery. We need some way to identify the right building in the dark."

"So, what," Alec pondered aloud, "like drop a flare on the roof?"

"I was thinking of something a little louder." While the Strike Eagle was groping blindly in the dark, we could see the target clearly. A Hellfire wasn't likely to have catastrophic impact on a building of that size but the impact at night sure as hell would make for a visible target.

"It'll spook everybody inside." Alec wasn't arguing, just running the numbers. "What kind of follow-up time are we looking at for round two?"

It was a good question, one that lacked a textbook answer. If Eagle One-Niner was on a solid approach, if he caught the flash right away . . . I pulled a number out of my ass. "Sixty seconds."

I heard Alec huff. It struck me as the sound of a man who had to run an unorthodox warfighter solution past a picket-line of bureaucrats and lawyers. I didn't envy him the task. Then again, I didn't cut him any slack either. "Better hurry, we're down to five minutes of fuel."

"You're a load of help," he groused before the phone clicked.

I briefed Joker and Leo on the possible plan so they could work the approach. Big was working vectors and windspeeds while Joker coordinated with Eagle One-Niner. Both planes veered into wide curves that would bring them each into their own alignments with the target.

I snatched the phone off the cradle before it completed its first ring. "Tell me something good."

I'd heard a great many things in Alec's voice since this effort started, from smart-ass snark to anger and frustration. This was something different. He sounded almost . . . happy.

"They said yes." Happy wasn't quite right, I realized. His tone was more like disbelief.

"Just like that? What did you tell them?"

The deadpan returned to his tone. "Do you really wanna know?"

I didn't need a moment to ponder that question. "No, actually, I don't. To be clear, we have a green light to engage the building?"

"You are cleared to engage."

"Put it in chat," I added as I hung up the phone, probably every bit as surprised as Alec. But by whatever rationale, the mission just stepped out of the endless vagaries of political agendas and into the stark black and white of Air Force combat—no convoluted policies, just physics. That was something I understood.

I turned to the command console. "Gentlemen, you are cleared to engage the building. Make sure Eagle One-Niner is in position and watching for the bright light."

For just an instant I saw the flash in the eyes of the flight crew, lightning that jumped along exchanged glances. Then everybody snapped to their respective tasks like a machine shoved into high gear.

"Eagle One-Niner, we will be marking your target . . ." I listened to Joker's rapid-fire directions. In my mind's eye, I could see the F-15E rolling into a steep bank, nose canted down, eyes wide.

The K-model Hellfire punched off the rail, accelerating past the speed of sound. Given the low chance of defeating the oversized structure, we held the Mike in reserve. My eyes were glued to the screen as I stood, silently counting the seconds.

Bang. The missile slammed into the roof, the blast effect more anemic than I had hoped. My gaze jumped between screens as I strained to listen to the voice of Eagle One-Niner.

"Wildfire Two-One, I have your target. Rolling in hot."

Yes. I didn't shout but sure as hell wanted to. I glanced at the clock, noting the red second hand sweeping past three. The Eagle would be coming in at a sprint.

On the thermal screen it was clear we had kicked the beehive. Glowing blobs poured out of the building, some wheeling around to look back while others had the common sense to simply run. Whatever it was that just happened, it would be a hell of a lot safer to figure out in the light of day—ideally from a few miles away. In times of war, a reliable fight-or-flight reflex is invaluable. The second hand was rolling past nine when Eagle One-Niner said, "Weapon away."

A cluster of blobs, a dozen or more, coalesced around a central figure and ran from the building, making a mad dash off for the treeline. They had covered some fifty or sixty yards as the second hand swept past twelve. The figures broke left along a dark line in the ground that might have been a low stone wall.

"C'mon, c'mon," I snarled under my breath, urging gravity to pull the bomb faster. The second hand was nearing two when— *ka-bam.* A starburst of white flashed from the center of the dark square, and for a moment the camera struggled to adjust to the moment of sunlight. Adobe walls buckled out, fragments of construction and bomb case cutting through the circle of stragglers. People, or pieces of them, were hurled outward in a cloud of debris.

Even with heavy ordnance like a GBU, lethality is never what you would expect. All it takes is the right intervening barrier, or a lucky dice-roll of fate, when the piece of frag that misses one rips the next guy in half. Some walk away from a blast seemingly unhurt, only to keel over—moments or maybe hours later—from the sledgehammer of overpressure that can cause the squishy parts inside a man to tear in awful ways.

While the F-15E streaked off into the night, we remained, staring down with an unblinking eye. Figures emerged from the wreckage, some wobbled an erratic path, some leaked streams of heat that left a glowing spatter trail in their wake.

As the survivors dispersed, so would our ability to track them all. We could pull the camera way back to cover a wider swatch of ground, but at the expense of detail. To see targets clearly enough to spot weapons or watch for specific actions,[73] we had to push the bleeding edge of our zoom lens. It was likened to staring down through a soda straw. If we lose track of someone, we lose that person for good—so we had to choose somebody. I glanced at the group huddled under the tree, squinting at the dot in the center of what remained a definable circle of protective bodies. *I vote for you.*

But tracking wasn't my call, and I called out to DJ. But it was Eric's voice that answered: "They say stay with the large organized group."

The flight crew glanced back for confirmation. I nodded, stabbing a finger at the screen where the white blobs had just begun to shuffle in short bursts down what looked to be a ditch or shallow gully. *The movement suggests a team trying to take cover from ground fire*, I thought with a grim nod, *but neither tree nor terrain will be worth a damn when shit falls out of the sky. Death from above.*

"That's our target."

The double-wide came up on voice and stated an AC-130 Spectre was in the vicinity.

The Spectre was a beast. From as far back as the 1950s, the hulking C-130 Hercules had earned its chops as a heavy-lift cargo craft. Driven by four massive turboprop engines, a "Herc" could carry some seventy-two thousand pounds of cargo. Early on in the Vietnam War, somebody posed the bright idea of hauling seventy-two thousand pounds of guns and ammo instead. What arose from Project Gunship II was a series of armed leviathans that culminated in today's AC-130H. If you imagine bolting wings on a tour bus full of guns, you are in the ballpark.

---

73 Sniper engagements share a similarity in that long-range marksmen observe targets through lenses at distances that often blur small details like rank insignia—details that might otherwise identify a senior officer as a priority target. But actions, like salutes thrown up by junior soldiers, can hang a neon SHOOT HIM FIRST sign over an officer that might otherwise blend in with everybody else.

The utility, which is to say the lethality, of the platform was enhanced by the mix of weapons. A GAU-12 Gatling gun could pour 25 mm armor-piercing rounds at a rate of about two thousand to four thousand rounds per minute. At that volume of fire, the tracers look like an unbroken stream of red laser burning down from the clouds. And that's the little gun in the lineup. In the middle position is the L/60 variant of the venerable Bofors gun, an automatic cannon that belches 40 mm high-explosive shells. Bringing up the rear of the lineup is a massive 105 mm howitzer. It trades rate of fire for a football-sized shell in a variety of flavors. In terms of oomph, a 105 mm can obliterate a main battle tank or kill a small building.

"Dragon Four-Four, this is Wildfire Two-One requesting . . ." Like clockwork, Joker rolled into the second engagement, swapping authorization codes and target locations. This time things clicked; given that only one building in sight had a smoking crater in its center, there was little question as the center point of our activity. Then it was a simple matter to walk Dragon on to the target.

I listened to the voices on the gunship, heard the sharp *chunk, chunk, chunk* over the radio just seconds before a series of white blossoms appeared around the base of the tree. Hot craters pocked the earth—more parts, more spatter.

I scanned the monitor. In the opening double-tap, some forty enemy combatants had been explosively whittled down to a dozen. That group was now reduced to just over a handful. Whatever team cohesion existed had gone up in the smoke that engulfed them. They scrambled haphazardly—every man for himself. If a protectee was still among them, the benefits of rank had expired. Now they just wanted to get the hell out of Dodge.

As they traveled down what looked like a dirt road, locals got out of their way, some off the road entirely. They had to be some dangerous hombres to be given such a wide berth.

The group came across a beat-to-shit piece of farm equipment which, by the glow of heat from the engine, was running. The makeshift tractor was dragging a hay cart full of villagers. A dog running circles around the cart was the cherry on top of the oddly comedic scene.

Our group surrounded the vehicle, we could see guns and arms waving madly. They forced the passengers off and clambered onboard, their escape then proceeding at a ridiculously slow pace. In disbelief, I watched the vehicle plod off, understanding why Hollywood, despite all of its silliness, has never filmed a chase scene using an old tractor as the getaway vehicle. That the dog kept pace trotting alongside simply emphasized the lunacy.

Chatter came over the radio; Dragon Four-Four was in demand. My brow knit as I looked across the screens, assessing our situation. The AC-130 was better suited to run down scattered stragglers on their own, while our remaining group had conveniently chosen to mass on a big piece of metal with an engine that was showing white-hot in infrared. Still, I hated letting go of an asset, especially one with that many guns. But we could handle the tractor.

"OK, guys," I said. Save for Joker at the stick, who was focused on flying, everybody else in the GCS glanced up. I pointed at the tractor and said, "This one is our job."

I couldn't help but think of what, in a Hollywood fiction, would have followed—an expression of bloodthirsty hunger or conflicted aversion. I saw neither in the eyes around the cramped room. Yeah, the guys on the screen were having a bad night, maybe the worst in their lives. But those guys were part of an effort to kill Americans, to bomb cafes and fly planes into buildings. It was our job to stop it. Nobody in the GCS took for granted the weight of ending human lives in a burst of fire and frag. But not a one of us shied away from that duty. All I saw was steely nods as the team set down to business.

I called Alec, taking advantage of the time it would take to work the engagement. The tractor was chugging along but had nowhere to go. We followed overhead, swinging a lazy figure-eight in the night sky.

"You got all this?" In hindsight, I saw it was a pointless question on my part.

"Yeah. We're working approval for a shot on the tractor." Alec and I had fallen into a synergy of our own, both of us working through our respective parts of the same equation. Knowing what was in his hand, I dealt him a few extra cards.

"Confidence of a hit is high; the thing is moving at a crawl. I know we didn't do any testing against a moving target, but our run in will take than into account. We'll approach straight down the road, giving us the best chance for success. If they stop, even better. They're in a pretty empty spot; collateral is low." All that translated in political-speak to "low chance of a career-ending fuck-up."

I was fairly confident that was the real make-or-break consideration on the policy side. But we had already cleared the hurdle of the opening shot, then doubled down by involving other assets. This was just closing out. If by some stroke of luck one of the key players was among those lucky enough to be clinging to the tractor, we had to see that his luck ran out.

It didn't take long for that to happen. For the second time tonight, we got the green light to deliver karma on the nose of a Hellfire.

I listened as Joker and Leo worked through the checklist: "Laser on . . . lasing . . . confirm lasing." The process ran like clockwork, the training, the diligence, all paying off. Joker gripped the stick, finger on the trigger as the countdown closed out in "3, 2, 1, rifle!"

The missile left the rail, punching out of camera view as it traversed the long arc to the ground. Leo kept the laser on target, the crosshairs fixed on the engine compartment that represented the vehicle's largest single mass. My eye caught sight of the dog and I winced. *Sorry pooch.*

As if he caught some ill-vibe on the air, the dog broke right and took off at a sprint. The guys on the tractor had no more than a heartbeat for that act of abandonment to register before the missile slammed home.

The impact was spectacular, a flash of thermal white that scattered chunks of hot metal and bad guys in one big infrared spray. Smoke rose from the center of the starburst as the dog trotted back, sniffing from spot to spot. We instantly named him Lucky.

Every eye in the GCS, the double-wide, and the GRC stared at their respective screens. We saw smoke, bits of heat speckles everywhere. Then something moved on the periphery of the scorch mark that had been the trailer. I squinted, then my eyes went wide. *You gotta be fucking kidding me.*

The figure must have been thrown ten feet from the blast, maybe fifteen. Yet he staggered to his feet, shaken but still mobile—by all appearance, relatively unharmed. We weren't chasing a terrorist, I realized. We were trying to kill the fucking Energizer Bunny.

And the Bunny's brought reinforcements. An uneasy twist ran through my gut as other figures converged on our survivor. They moved with a purpose, not the kind of organized action that a happenstance mob will do right after something just exploded. Some of them were picking up weapons from amid the wreckage as others looked to be assessing possible cover. Sure enough, they headed for the closest structure with our last survivor in tow.

The building was small, from overhead a narrow rectangle reminiscent of a single-wide trailer. The survivor, along with his newfound allies, hustled inside. I had a mental image of hands frantically pulling drapes shut. But for all their fear, we circled overhead with two empty missile racks. On our own, we couldn't drop so much as an insult.

But while calling in yet another asset was an option, we faced the possibility of yet another laser conflict, followed by another talk-on, this one presumably from the wreck of the tractor. Even that had become more complicated; the number of burning holes in the earth had increased, adding confusion to voice-based targeting.

Fatigue was another factor. While Joker was running solid, Leo had been eyeballing the screen long enough to merit a break. Will B. was the next sensor operator in the queue and swapped seats with Leo as we worked through our choices in air assets. He was one of our younger guys who had come in on the second wave. He was about to get thrown into the deep end of the pool in a hurry.

The cherry on top was fuel. Our reserves really were burning slowly closer down to my predicted five minutes before we'd have to head home. Once that happened, there would be no time for us to adjust for another miraculous escape. Whatever we threw next would have to be definitive—a game-ender.

I waited while the pilot and double-wide worked our various communications means to see who could come play with us. As much as I love my Air Force, the answer in this case would come courtesy of the Marines. I got Alec on the phone.

"We've got a Marine F/A-18 with a pair of AGM-65 Mavericks."

Alec was unfamiliar with the designation and knew terminal effects would carry a lot of weight with decision-makers. "Will that get the job done?"

"Get it done?" I chuffed. A G-model Maverick packed a 300-pound blast/frag warhead designed to kill buildings. The business end alone weighed over three times the entire weight of a Hellfire. "Oh yeah. Trust me on that."

As our fuel gauge continued to dwindle, we vectored the F/A-18 on target. It was a long shot, and Will had to hold the laser rock-steady on the wall as the missile tracked in. It was a perfect hold. The building disintegrated from right to left as if swept off the map by a giant hand. The violence of impact was far greater than anything we'd grown accustomed to with Hellfire. A low whistle drifted from somewhere in the GCS.

For a long several minutes we stared down from the night, watching for movement, looking for any sign that something had survived—a flipper, a squirter. There was none.

I looked at the clock. Some two hours had passed since we first rolled in over Tarin Kowt. To the best we could tell, none of the forty or so enemy combatants had survived the night. Our work done, we banked north and headed home.

14: THEY GROW UP SO FAST

ALEC BIERBAUER

WHEN YOU GET THE ANSWER, THEY CHANGE THE QUESTION
October–November 2001

Throughout the life of this effort, problems would arise in an infinite assortment of flavors. By now we had become pretty resilient to the sudden appearance of oddball obstacles, technical challenges, or bad luck. But nothing in our limited bag of experience-to-date prepared us for a démarche by the Russians.

Mind you, discovering that the Russians were even aware of our top-secret project was unsettling. Most of my mentors in the Agency had cut their intel teeth in Cold War struggles against the Soviet empire. Almost ten years had passed since the USSR had collapsed, but the phoenix that was modern Russia looked and acted a whole lot like the same old bird with a few new feathers. Having that bird stick its nose in our business was every bit as unwelcome.

No matter how covert we kept our airfield and personnel, a measure of global attention was inevitable when Hellfire missiles inexplicably started raining down out of the sky. Al Jazeera speculated on some sort of new Western capability. The Russians, having little in the way of specifics on which to hang their *ushanka*, took a wide blind swipe at the way missiles are bolted onto aircraft in general. We received word through diplomatic channels that we were not allowed to hang missiles under the wing of nonmilitary aircraft and that doing so was a violation of the cruise missile treaty. Different than a cruise missile, the armed but unmanned Predator could return home without being detonated, thus creating a sizable difference from a cruise missile that we would not want to return home in a mission-abort scenario. Nevertheless, we took the objection seriously.

Conventional wisdom would state that this would put a real cramp in the ability to conduct flight operations, almost crippling our ability to respond quickly to events. Prior to flight, a missile has to be connected to the aircraft and run through a series of electronic tests that insure proper connectivity and function. With most aircraft, none of that can begin until we hang the missile on the underwing pylon—most aircraft, but not Predator.

Whereas the intent of the regulation was arguably clear, a quality Western education had taught us that the power of any law, certainly the teeth, was in the wording. In this case the word "hang" meant to suspend from above. As it turns out, one of the unexpected benefits of Predator was that we could unbolt the wings, flip them over and *rest* a missile on top of its pylon while doing all normal testing. Satisfied with our explanation, America dutifully and honestly responded that we were in full compliance with all relevant treaties and politely suggested that Russia *otvali*![74] In the brief game of Our Lawyers Can Beat Up Your Lawyers, the final score was USA: 1, Russia: 0.

Mission demands for Predator rapidly expanded outside of all initial planning, and people who mocked us yesterday now needed our help—desperately. Additional roles, like close air support for ground-rescue missions, suddenly became real-world mission

---

74 Piss off!

requirements. The game changed from just killing bad guys to keeping good guys alive, which we were eager to support.

In the realm of the bizarre was an event that involved the improbable trio of Jim Ritchie, a call to the GRC, and Abdul Haq, a one-legged Afghan warlord.

During what had become an otherwise benign shift in the GRC, a call was received on an unsecure line on the desk used by several persons, but primarily the USCENTCOM liaison officer, Lieutenant Colonel Richard P. He was not at his desk at the time and was likely taking a break from the abuse of the mission managers and staff in the GRC. Richard had the unenviable task of interfacing between the CIA and USCENTCOM for our operations. In most cases, USCENTCOM didn't share our desire to move swiftly, and he was often caught between two competing bureaucracies. It was not uncommon for him to find labels plastered across his computer with cute motivational sayings like "Put down the Barbies."

On the line was a highly anxious person who identified himself as "Jim." Jim stated that he was calling on a satellite phone in Peshawar, Pakistan, and that a "friend" was in desperate trouble and needed our immediate assistance.

Keep in mind, at the time satellite phones in that region were primarily used by DOD ground forces, CIA officers, CIA assets (indigenous sources), and those others of sufficient means to afford the up to seven-dollar-a-minute call charges. To further narrow the field, very few people within the CIA had the direct-line number to the GRC, and precious fewer yet outside the CIA. So when this call came in, it was thought to be from one of our case officers in the field—effectively, an epic 911 call for help.

Taking the moment further into the bizarre, the caller indicated Abdul Haq had crossed into Afghanistan with a small force from Pakistan and had come into contact with a Taliban element. Abdul Haq was a one-legged Taliban opposition figure who, much like

Hamid Karzai, hoped to be the leader in a new Afghan government. Haq was known inside our intelligence circles and would be one of a number of supportable opposition figures favorable to US interests.

Using Jim as a relay, the mission manager was able to query for the specific location of Haq and his associates, the location of the pursuing Taliban, and the available weapons, equipment, and plan of Haq if he could clear his pursuers.

During the discussion, Jim had Haq alternating between hiding in the scrub brush and trying to move to create distance from his Taliban pursuers. Jim continued to plead for assistance to Haq, and the mission manager told him we were considering weapons options to help disrupt the Taliban pursuit but that we had limited firepower and Haq would need to be prepared to fend for himself. In fact, we were considering the use of the two on-board Hellfire missiles along with coordination with any available assets for further assistance.

At this point, the mission manager began to find the dialogue with Jim as odd. As the mission profile to engage the Taliban was being established, a parallel engagement sought to establish if Jim was in fact a CIA, DOD, or even a government employee and how things had progressed to their current state.

Using the chain of command, the CIA made the determination it was a valid target to engage the Taliban element in order to help protect Abdul Haq as a potentially viable figure for the future of Afghanistan. By that point in the conversation, Haq was hiding in some brush while a truckload of Taliban was circling in close proximity.

During subsequent satellite phone calls, we determined that Jim had a last name, Ritchie, and was a private citizen promoting Abdul Haq. Jim and his brother, the kids of missionary parents, had grown up working in Pakistan, and they understood the landscape well. As adults, the Ritchie brothers became wealthy do-gooders and saw the aftermath of 9/11 as an opportunity to promote a stabilizing influence in Afghanistan. Unfortunately, they neglected to properly coordinate their endeavor and failed to adequately equip or secure their candidate when they sent him across the border armed with a pistol, a crutch, and two or three well-intentioned guards.

Throughout the hour or so that the engagement was being discussed and developed it became clear Jim was not operating under the direction of the government. The next logical line of questioning was to try and determine how Jim got the phone number to the mission manager for a classified program. Probably, one of a number of well-intentioned, properly cleared visitors from either a congressional oversight office or a White House staff made it available after any one of a number of briefings given to the small circle of witting individuals who visited the GRC.

The decision was made to engage the vehicle. The shot was aligned and permission to engage was granted. Within minutes, the threat posed by the Taliban vehicle ceased to exist.

The second missile followed against a set of dismounted Taliban fighters. Tragically, the Taliban had brought more guys than we could kill with just two missiles.

In the end, all Predator could do was watch as Abdul Haq was ultimately captured. In the subsequent days he was transported to his hometown of Jalalabad where he was tortured, killed, and put on display as a warning to others. It was a grim reminder that despite the remote nature of our war, the need for success, as in any war, was still measured in flesh and bone.

While all this was going on, arm wrestling between priorities began to heat up—"no-shit ops" against the proactive collection of intelligence. Depending on what agency one talked to, both were mission critical. It raised the pivotal question: had the arming of Predator made it more valuable as a weapon than as a spy? Coordinating all the three-letter players on both political and practical levels became a nightmare.

Yet another mission set included exploring the use of Predator to validate intelligence assets. These were often well-placed people whom our spy network had manipulated into betraying their nation. Predator allowed us to validate critical points without exposing human collectors to unreasonable, or outright suicidal, levels of risk. We also expanded the use of Predator to vet intelligence anecdotes, debunking assertions that terrorist training facilities were such things as NGO grounds or even a purported "falconry camp."

But too much good all at once can be a bad thing, and the advent of real-time global streaming from Predator signaled the potential breakdown of prior intel-management protocols. We could now capture SIGINT from Predator with our crude adaptation of a communications radio. Suddenly Predator was ingesting signals data, pushing unfiltered information directly to other intel agencies.

This led to cries for architecture-design revisions and collection-management policies. With the prospect of a truly epic turf war in play, the "one team/one mission" effort devolved into a bristling furball of cats and dogs!

OH, THAT'LL LEAVE A MARK
November–December 2001

After roughly a dozen Hellfire engagements against al-Qaeda and the Taliban, we noticed a disturbing pattern. The targets were mostly people in the open, thin-skin vehicles like Toyota Hilux pickup trucks, and small to midsize structures. Our weapon was originally designed to penetrate tanks and other armored vehicles.

On the positive side, Hellfire was hitting targets with amazing precision. You could count on a direct hit on the sensor operator's laser spot within about thirty seconds of pulling the trigger. The operators were getting so good and confident with the weapon that our biggest choice on shots was to put it through either a window or the front door. I thought it was bravado when, through my open phone line with Mark in the GCS, I heard Gunny, the sensor operator, ask which window we would like a specific shot to go through.

"Mark, is he serious? We just need to hit the building on this one." We were addressing a very large structure called Bagh-e-Balah that we knew was full of bad guys at a meeting. Hellfire was way too small to deal with the attendees in such a large structure with any degree of confidence. With no alternative weapons platforms to bring in for a bigger strike, Gunny put it through a window high on an ornamental dome to the building. The intent was to flush key persons from the meeting and strike them if possible in their vehicles or in the open.

The strike was spot on, and as expected the building was vacated in a hurry. We found exactly what we wanted, what we had hoped for in the Omar scenario on the opening night of the war. A diamond security formation, surrounding a small group of principals, moved briskly to a three-vehicle convoy outside the structure. We couldn't get a second shot lined up on the vehicles before they started moving from the secluded hilltop venue toward the busy streets of Kabul.

The three-vehicle convoy sped away from the building and stopped about a half mile away just before hitting the public streets. We lined up a shot on the middle vehicle that had the occupants of the first and third vehicle gathered around it. The strike was authorized and, as expected, thirty seconds after leaving the rail it was a direct hit—center mass on the middle vehicle. Not counting the occupants of the vehicle, six or seven people stood within six feet of center mass and five or six of them were seen scrambling from the scene when the blast cleared—not flippers, but squirters. This was the most blatant demonstration of targets surviving a shot, likely as a result of our weapon overpenetrating a thin-skinned vehicle.

Just then my phone blew up. My bosses were on one line, rolling out their frustration with our effects on targets, when Mark called me on another line. Before he could get a word out as to why he was calling, I jumped in with "Mark, I got the boss in my other ear asking why your weapon seems anemic. I will let him know you are working an assessment."

"For the record," Mark snarled, "Army weapons are the issue here. You may have noticed that Gunny has no problem delivering them on time and under budget."

"Fair point," I conceded. "You got anybody you can ping on that one?"

Mark came back with a heavy dollop of snark, "Sure, lemme check that for ya," before he snapped abruptly back to mission. "Hey, the squirters are loading the first vehicle and on the move. I assume you want me to follow them?"

"Follow that car; let's see where they go. We can address any new targets that develop. These are known bad guys—keep eyes on them!"

The vehicle turned into heavy downtown Kabul traffic and worked its way through town for about ten minutes before pulling up in front of a residential house that was nowhere on our radar as a bad-guy location. We had kept custody of these guys since back when Gunny put a hole in their window to break up the meeting. We sat on the structure but were out of weapons, and with the residential nature of the area with houses almost on top of each other, we had nothing on the USCENTCOM published air tasking order for the day to indicate there were aircraft to help manage a low-collateral-target weapon option.

We stayed on target with our remaining hang time and had another bird with two more Hellfires on its way to relieve us in a couple hours. With a fresh bird on station we looked at engagement options using our Hellfires. The decision was made to engage the target. Ever since Gunny had shown off his precision strike skills, that level of thread-the-needle performance had become the expectation.

"DJ, get Mark on the phone and tell him we are cleared to strike the house. NIMA thinks putting it through the upstairs window will have best effect." Mark had been anticipating the decision and was already down at engaging altitude, with a dry run under his belt. The strike was textbook, and the effects on upstairs of the fairly small structure were impressive; that said, three or four people exited the front door into the street.

I struggled to keep my tone flat and professional. "Mark, you are clear to engage the personnel in the open who exited the structure."

"Roger, standby," Mark replied, before adding, "put it in chat."

The next missile came off the rail, and after the normal one or two seconds of a blacked-out image the next thing we saw was not the target in the crosshairs. Something had gone wrong, and the laser now pointed on some building randomly selected by Murphy's Law, with the missile streaking in that direction.

"Mark, what the hell happened to the target! What are we looking at!"

"Standby, Alec" Mark said calmly.

"Mark!"

"Standby," came the calm reply.

While I waited for the longest twenty-eight seconds of my life, the sensor operator calmly went to a wide screen, identified the right neighborhood and then the right street and the right house before carefully adjusting the laser designator back to the right house. The laser settled in on the front door less than two seconds before the missile arrived.

"Yes, Alec, what do you need?"

"Never mind."

The net effect of the four missiles shot over a fifteen-hour period against the group meeting at Bagh-e-Balah and their subsequent locations resulted in a significant impact to al-Qaeda and their Taliban hosts in Kabul. Although the ability to engage a target and maintain pursuit was an amazing development as was demonstrated earlier in Tarin Kowt, it should not take four $100,000 missiles and over fifteen hours to successfully engage that type of target. Redstone gave us answers, recommendations, and quick solutions.

The response was equally impressive to their initial effort on Hellfire. Chuck came to DC with nothing more than a cocktail napkin sketch of a quick modification that would enhance lethality.

The engineers at Redstone felt that since this was not being put on a manned platform it could dispense with many of the test and evaluation efforts that would be required if manipulating a weapon located in close proximity to the pilot. Since our pilot was about seven thousand miles away from the plane he was flying, we agreed. Together with the cost and schedule savings it just made good sense. We needed a quick solution and didn't have much money or time to put into it.

The result was a devious solution that involved a heavy metal called tantalum. Chuck and the engineers came up with the concept to wrap a tantalum sleeve with a Pearson-notch laser cut to maximize the number of appropriately sized fragments to be added to the effects of the Hellfire on impact. In order to not interfere with the flight profile of the Hellfire, the sleeve would be hinged and wrapped around the center of gravity of the missile. Once Didi was able to model it, the effects were clear: a significantly enhanced fragmentation panel radiating out from the missile body to disable soft targets.

The entire process, from us defining the problem to Chuck doing basic field mods with super glue on our deployed missiles, was about six weeks. That was another outstanding accomplishment.

15: WITH EVERYTHING ON THE LINE

MARK COOTER

WILDFIRE
November 2001–March 2002

By all expectation, November 13 had the makings of a quiet night. We were in our groove. While we were in the night sky high over Kabul, we had no critical tasking, just more target development.

Circling over the city, we worked pattern-of-life collection. It was the kind of night I least expected to hear from Alec. I picked up the phone to be hit with an immediate question.

"How quickly can you put us over the Gardez Market Circle?" Alec's voice had an edge of urgency.

I glanced at the tracker display to see our location. Gardez Market Circle was less than a mile from our current position. I didn't need pilots or calculators on this one. "Five minutes," I replied.

Alec must have put his hand over the phone. I heard a muffled exchange, and while I couldn't make out the words it was clear the other voices were equally charged. Alec came back just a moment later. "Okay, let's roll on Gardez Market Circle."

"I wish to hell you guys could make up your minds."

Alec bristled, barely a word into reply before he muffled the phone once again, barking at somebody on his end of the line. He came back with the tone of someone who had insider info on a big stock market move. "We may have Atef."

Whatever snark I had harbored to that point vanished at the word "Atef." Rarely does somebody start out in life with a goal to become an international terrorist. Most truly epic bad guys can trace their lineage back to some humble, perhaps even normal, origin. A scant few might even have been considered "good guys" at some point before turning horribly evil. Such was the twisted road of Mohammed Atef al-Masri.

After a short stint in the Egyptian Air Force, Atef assumed the quiet life of an agricultural engineer, with a sideline as a cop. But something happened along the way, a fork in his life road that led to fanaticism. Joining al-Jihad, a group of Egyptian Islamic militants, Atef traded fields of greens for rocky soil sown with land mines.

Operating out of Peshawar, in the Khyber province of Pakistan, Atef cut his combat teeth fighting the Soviets in Afghanistan. At al-Qaeda's Khalden training camp, a vocational school for terrorists, Atef befriended Ayman al-Zawahiri, who was implicated in a number of atrocities from bombings to the assassination of Pakistan's former Prime Minister Benazir Bhutto. Al-Zawahiri was among the elite in al-Qaeda; from there, it was but a short step to connect directly with Usama bin Laden.

When playing games of war in the Middle East, there are a lot of ways to die. Drowning doesn't usually make the list. So it was surprising when the *MV Bukoba*, a Tanzanian ferry, sank in Lake Victoria back in 1996. Of nearly a thousand people who perished in the sinking was Abu Ubaidah al-Banshiri, then second in command of al-Qaeda and bin Laden's right hand. While roundly hailed as a tragedy within the Islamic cause, for Atef this was the door of opportunity.

It was not a smooth transition. Atef became a lightning rod for criticism from fellow radicals, prompting none less than bin Laden himself to quash the infighting and bestow upon Atef his personal approval. On the strength of that endorsement, Mohammed Atef al-Masri became the military chief of al-Qaeda.

He took to his task with a bloody zeal, helping to lead the simultaneous bombing of US embassies in Tanzania and Kenya in 1998. Despite a $5 million bounty on his head, his next piece of work was the bombing of the USS *Cole*. In the blink of an eye, I was back outside the gate at Ramstein, watching what seemed like an endless row of flag-draped coffins roll by.

My unspoken gut response was *You're shitting me*, but for whatever mischief lurked in Alec's sense of humor, he wouldn't kid around about this. My fingers tightened on the phone, a grip fueled by the combined buzz of anger and adrenaline. "What have you got?"

"SIGINT caught a call from a phone linked to someone perpetually close to Atef. The caller described their position as fifteen minutes out, heading into town on the Gazni Line. If we're right, the vehicle will enter Gardez Market Circle from the northwest, then break away on the second right."

Fucking patterns of life, I thought with a wry chuff. All those mind-numbing hours circling overhead to document the coming and going of bad guys, of the friends and families of bad guys. All the analysts building maps and timetables, for a moment just like this. I shook my head, vowing never again to bitch about Circles of Boredom.

"We're on it" I said briskly, my mind shifting to task. "I'll hit you when we're over the target."

I quietly set the phone back on the cradle, consciously pushing the image of hearses out of my mind. Tonight ceased to be just another mission; with a bit of luck, it would become a reckoning. I glanced up at the feed from Predator's nose and watched headlights crawl like glowing ants along ghost-gray roadways. One of those ants was Mohammed Atef, and tonight he was going to die.

We rolled on station over Gardez Market Circle and settled into a slow counterclockwise loop, banked slightly so the camera naturally stayed on target with a minimal need for human adjustment. From our vantage point, we could track every route in and out of the circle, extending out a couple of miles in all directions. Even in the dead of night, cars and trucks rumbled along their way, not in the hectic bumper-to-bumper madness of the business day when major roadways can turn into parking lots, but hardly empty.

Despite having what amounted to box seats over the fifty-yard line, picking a specific car out of traffic on little more than heat was no small trick. Given the tendency for a heavily armed entourage, senior terrorists often favored SUVs like a Land Rover over smaller, more restrictive sedans. Being higher up the food chain carried the need for even more protection, the single Land Rover turning into a column of two or more. If the entourage held true to form, we would be looking for two or three SUVs rolling into the circle from the west, taking the second right, and heading out of the circle roughly to the southeast.

Half a dozen "possibles" had entered the circle, vehicles of the proper type or groups that looked to be running in a convoy, only to split off in different directions or take a wrong exit from the circle.

"Batter up." At sensor, Ken Mitchell's voice was calm, noting the next possible hit with neither excitement nor boredom. "Single Land Rover. No chase cars, but he's moving briskly." Ken glanced up. "Like a man with somewhere to be."

Over chat came "We're getting comms over the ARC-210 that someone is approaching the circle and headed to the meeting." Cliffy's $12.95 mod may just help lead us to Atef.

I watched as the light gray rectangle swept around the curve, passing the first exit, shifting left one lane and looking to pass the second. *Damn.*

At the last moment the vehicle shifted right, crossing the intervening lane and swinging off Gardez Market Circle. It was the kind of lurching move one might expect from some idiot texting while driving. My eyes narrowed. Or somebody was trying to make sure he wasn't being tailed.

That little bit of tradecraft might shake a Hilux off a bumper, but it didn't mean shit to an eyeball parked twenty thousand feet overhead. Ken kept the crosshairs centered on the Land Rover through the next left and watched as the vehicle made its way deeper into the city.

It pulled to a halt in front of a sizeable compound. The vehicle had barely come to a stop when an ant line of figures streamed out of the closest building. The dots surrounded the vehicle, others taking position to form a haphazard but definable perimeter.

The passengers climbed out of the Land Rover, their individual thermal signatures merging with the group that collapsed on them. As a mass they disappeared into the building.

Replaying details in my mind, I spoke to DJ. "Get Alec on the line."

A moment later, Alec was on the other end. He was amped: no hello, no yanking my chain for the fun of it. Instead he opened with, "We need this one, Mark," adding just a moment later, "This is big."

Well, thank God we've got the CIA to clear that one up for us, Captain Obvious. I bit back my natural response, restrained by the weight of what was playing out. I shifted instead to the analytic. "How big?"

As often his way, Alec answered a question with one of his own. "What's the biggest package you can call down on this place?"

That one caught me by surprise. In most engagements, 90 percent of our time is spent getting permission to shoot anything at all, ruling out ways that some innocent bystander might get hurt or that some window on a mosque might get broken. "How hard can we hit them?" was a new question, and not one to ask the US Air Force. God may have invented smiting things from on high, but the Air Force made it a science.

"Stand by," I said.

I called to the double-wide. "What other assets are available?" We had a lot of aircraft in the battle space, but fewer were sent into the higher-risk volume of civilian airspace. But we had choices: a beefy F-15E Strike Eagle, an F-16, and an AC-130 gunship lumbering along out over the foothills to the south.

The compound sat in a heavily congested part of town, with other structures on every side. Missing our target wasn't my concern. As far back as Iraq we'd been flying precision-guided weapons through specific windows. My concern here was what could happen when we hit this building dead-on. A two-thousand-pound JDAM would level the structure, and many more.

Detonating a literal ton of Tritonal, an explosive that packs about 18 percent more wallop than TNT, will turn a building into a multistory hand grenade. Parts of the target, anything from stone or concrete to furniture and appliances, can be punched into the house next door at a couple times the speed of sound. At two thousand pounds, PK for the neighbors is pretty rough as well.

"I don't think we can go over five hundred." I'd been in this business long enough to have a sense of effect just looking at a scene. As our viewpoint slowly orbited the target, I ticked through the weapons in our inventory, doing from-the-hip calculations of weapons solutions and collateral-damage estimates. I mentally sketched concentric circles that radiated out from "obliterated" through "fucked up pretty bad" on the way to "probably survives."

As computer programs go, Cooter 2.0 might lack the sophistication of an arsenal of multimillion-dollar Joint Service weaponeering and collateral damage tools, but there was no faulting it for historic accuracy. The five-hundred-pound Mk 82 was an Air Force standby that traced back to the 1950s, long before we bolted nose kits that gave dumb iron bombs a laser-guided precision. By now we had more experience dropping 82s across the planet, in every imaginable condition, than just about any other weapon.

"Yeah," I concurred with myself, "An Mk 82 is the right answer."

The double-wide reached out to the CAOC and requested assistance, which came in the form of a flight of F-15s. There is no other way to put it: the Eagle is a badass. With hundreds of engagements to its credit, the F-15 has never lost a dogfight—ever. That's a hell of a record. These Eagles were E-models, designed for air-to-ground attack, and they were damn good at that job as well. The beefy turbofan engines in an Eagle could drive it like a rocket through fifty thousand feet of climb in one minute, or flat-out to a speed on the order of 1,900 miles per hour—about two-and-a-half times the speed of sound.

That alone might sound scary, but in combat terms a plane is measured by the punch it brings to the fight, and that's where the F-15E becomes downright terrifying. Setting aside the 20 mm Vulcan cannon nestled in its airframe, the Strike Eagle can carry roughly twenty-three thousand pounds of external fuel and weapons underwing. Setting aside maybe five hundred pounds for a LANTIRN and another couple thousand pounds for air-to-air missiles, we are still left with a veritable hardware store of ordnance choices—delivered hot, or the next one is free.

We kept watch on the target as the F-15Es were vectored into our location. When dependent on somebody else to deliver the goods, the biggest fear is having the intended recipient leave before the package arrives.

Although capable of bullet speeds, the incoming F-15Es were coming in way below subsonic, avoiding the tell-tale thunderclap of a sonic boom that would alert the world to their presence. At the top speed of a Boeing 747, say 600 mph, our Eagles weren't coming at a run; hell, for them it was barely a jog.

We took the time to exchange our laser codes, the highly sensitive pattern of pulses that identify one point of laser light from all others. The last thing we want is a targeting system that can inadvertently lock onto another laser pointer by mistake.

I looked at the clock, once again running through my ever-changing mental checklist: good approach, targets all tucked inside; no visible collateral, like the Hollywood school bus full of orphans pulling up unexpectedly. It would be a tight shot, but just about as clean as we could ask for.

Unfortunately, as it had on the night of the tractor shot, the AN/AAQ-14 targeting pod in the belly of the Eagle proved to be a five-hundred-pound, $3 million piece of shit, at least as far as we were concerned. Whereas the newer targeting pod used by our Air Reserve Component, aptly named LITENING, could see a blazing laser dot in the dead of night, the LANTIRN in the F-15 could not.

Joker was sitting in the center-seat position between Troy and Ken, a dry-erase board perched on one knee. His expertise in combat comms was about to be the difference between success or failure. "Clock's ticking. Joker, talk him on." I snapped the direction, feeling any sense of margin quickly evaporating.

A talk-on relied on verbal descriptions starting from an inescapably obvious start point. If you were to describe the location of the mouse on your desk you might say, "about eight inches to the right of the keyboard." Units of measurement are considerably iffier in close air support, especially at night in varied terrain. What looks like a mile to one guy might appear to be half that to the next. The trick was to call out something in the area as a unit of measurement. In our desk example, a coffee mug on the desk might make for a decent "ruler": "The mouse is about three coffee cups to the right of the keyboard." It was old school, but it was all we had.

Joker picked out a soccer field. A couple lights burned at opposing corners, making it fairly visible. He declared a soccer field as his unit of measurement.

"From the large traffic circle," Joker said firmly, articulating every syllable. On the whiteboard, a scrawled arrow indicated north, as various waypoints appeared one after another. "Target is four soccer fields away at a heading of two-seven-five degrees."

Raven Zero-Six, the lead Strike Eagle, would repeat the direction, several moments passing before responding with a new form of uncertainty.

In all fairness, the Strike Eagle pilot had just a few moments to pick out buildings in the dark while smoking by at several hundred miles an hour. Joker continued to refine the verbal picture, providing cross-references from various angles. "OK, it is two soccer fields due west of a large courtyard with a fountain in the center."

My eye flicked to the minute hand that had crawled through some sixty degrees of arc. *Cripes, a sixth grader could have figured it out by now.* At this rate, the Strike Eagle would be out of fuel before it stumbled onto the right address. Just four short minutes later, Raven Zero-Six declared bingo fuel—out of gas. With barely a "sorry 'bout that," the Strike Eagle banked left and headed back to base.

"Fuck this, let's see if we can get some Marines." I wear Air Force blue and do so proudly, but the Marine F/A-18s had proven themselves able to interface with Predator in the past.

Close air support was a staple for Marine pilots, and they came in hot. We exchanged laser codes, compressing the step-by-step into more of a burst transmission, certain that at any moment

Atef and his minions would load up and disappear. The Hornet rolled in on target, confirming sight of the laser and accurately describing features to either side. We were all looking at the same building.

Of course, nothing is ever that simple. We can't fix one problem without absorbing another. This new one was less of a shortcoming than a problem of abundance. Unlike the F-15E that carried a Chinese menu of choices, the F/A-18 had but one flavor of laser-guided bomb, the GBU-16, a Mk 83 class weapon. To the layman, one digit up from an Mk 82 might seem like a trivial matter, but the GBU-16 Paveway II weighed in at a hefty one thousand pounds, twice that of the Mk 82.

I winced. That much high explosive pushed the PK in the target building above a perfect "one," or as we used to say in the south, "It'll kill you and the horse you rode in on." From a technical perspective, it also meant that things could get a little frisky at the neighbor's house, at least on the sides facing our target.

I spoke into the phone to the GRC, surprising myself with my understatement. "Alec, we are good to drop. Confidence of a kill is high."

I heard a muffled exchange between Alec, the mission manager, and others I couldn't recognize. After a long moment, Alec came back on the line. "I just got the word; you are cleared to drop."

I clicked off the line and nodded to Joker, who relayed the command to the Hornet. "Call your release."

At a bit more than ten thousand feet in the air, the GBU-16 Paveway II detached from the underbelly of the Hornet, falling into a long graceful arc like a Hail Mary pass in football. It accelerated as it fell, hitting a terminal velocity of just about the speed of sound before it slammed into the roof of the compound. The fuse in the nose held off just a couple milliseconds, allowing the mass of the bomb to punch fully inside the structure before detonating.

For a moment the entire infrared video stream went white. Then the system compensated for the sudden bloom, darkness returning to define a fireball that engulfed the structure. Chunks of masonry shot off in all directions.

I stared at the image as the fireball consumed itself, the roiling volume of energy collapsing on its center to abruptly fall back into darkness. In just moments the scene fell still, revealing details in what we could see and what we couldn't.

We could see a structure that had largely pancaked into the ground. I'd done enough BDA (breakdown analysis) to know lethal effects when I saw them. But what we didn't see was every bit as important—no fires in adjacent buildings, no structural collapse outside of the target proper. And no one exited the compound, not runners or walkers, no squirters or flippers. We had brought down the proverbial hammer of Thor, with precision and finality.

Staring at the smoking hole, I felt a moment of closure. The thought that ran through my mind was worthy of the cowboy hat that hung on the wall: *That was for the* Cole, *you son of a bitch*.

WHEN IT ALL HITS HOME
December 7, 2001

War is a business of fighting, of hunting and killing the enemies of the nation, of protecting family, friends, and neighbors from a distant enemy. Despite anything people say to the contrary, war is sometimes about seeking retribution for lives taken. I didn't personally know a soul that died onboard the USS *Cole*, but I didn't need to know them to care. They came home draped in my flag, under which they served to their final breath, a flag they felt worth the sacrifice. No heart isn't crushed by such a terrible loss. Brothers and sisters I'd never met had died protecting my family, friends, and everyone else back home.

As we settled into our rhythm in the battle space, we found ourselves increasingly able to sidestep into the unknown, to bring utility to a fight where none existed a moment before. By the very nature of our existence we could appear as if by magic in a piece of sky—an angel to avenge in one moment or to protect in the next.

But that existence, again by its very nature, was somehow separate from the battle we so easily influenced. We could see the fight, often far more clearly than the flight crews who suffered firsthand the slings and arrows of antiaircraft fire or knew the terror of a missile streaking toward them. And while every day we used phrases like "we could die" or "we could crash" the truth of it was that "we" could not. Should our plane corkscrew into a mountainside, or be blown from the sky in burning shreds, we would all stand up, shut down our systems, and go home. We'd live to fight another day.

For many of us this was surreal—a dichotomy without precedent. To anyone who has risked skin in combat, who knew the lonely chill of crossing into enemy territory, war and risk were inseparable. A soldier could perish or maybe the one next to him or her, but one way or another that risk was always very personal. I think we all processed the chronic sense of detachment in different ways, right up to the moment that war made it personal again.

For Alec, that day came just the prior week when news came in of a Taliban uprising at the Qala-i-Jangi fortress near Mazari-i-Sharif in northern Afghanistan. Mike Spann had been in that fight, a colleague of Alec who served in the CIA's Special Activities Division. Spann held his ground against insurmountable odds, killing several before his pistol ran dry and perhaps more with his bare hands before he fell beneath the crush.

I knew it had hit Alec hard. News of the loss tore not only through the Agency but also throughout the whole of the Special Operations community, where Mike had amassed a reputation as both a leader and warrior.

While I had hurt for Alec those past days, those words seemed suddenly dull and distant as I now sat on the edge of my bunk with a piece of paper in my hand, which, very uncharacteristically, trembled.

"Two American and twenty Afghan casualties." The words on the page carried that clinical detachment you can only find in a news report. A list of names appeared, divided between the wounded and the KIA—Killed in Action. There, amid what to me were faceless Dans and Brians, was one Master Sergeant Jefferson Donald Davis. I knew him as Donnie, and I knew his face like I knew my own.

Donnie and I went to high school and played football together in the small town of Elizabethton, Tennessee, home of the fighting Cyclones. Our friendship strengthened throughout college and continued when we both went off to the military.

A life, the report said, that was cut short when Donnie's team, a Special Forces group out of Fort Campbell, was waging a last-stand battle against Taliban forces near an enemy stronghold north of Kandahar. In the fog of war, air support intended to provide much-needed firepower went astray. A two-thousand-pound bomb intended for the Taliban fighters slammed into the ground just a hundred yards from the clustered Americans. To a bomb that size, a hundred yards is nothing.

I dressed and went to work in a haze, passing through the double-wide in silence. Punching up the internet, I found a photo of Donnie and printed it, then headed for the briefing room. The proceeding ran as usual, with the same well-oiled routine: weather, intel update, mission taskings, and the obligatory "other stuff" tacked on at the end. Briefings typically closed with some words of wisdom or encouragement usually from Colonel Boyle or me.

Try as I might to hold back the tears, I stuck Donnie's picture on the wall and turned to the room with glistening eyes.

"This is Master Sergeant Donnie Davis," I began, certain my voice was going to break, "my high school classmate and my friend." The room fell to silence.

"He was killed yesterday, along with twenty-one others." I swallowed hard, struggling to hold it together. "If we'd been there, if we had been overhead, we could have prevented this from happening. We must . . ." the resolve cracked, "we absolutely must do everything we can to support the guys on the ground." I turned and walked out as quickly as I could.

THE BATTLE OF ROBERT'S RIDGE
March 4, 2002

"What have you got?" Calls between Alec and I were now always direct. I was in the double-wide, and we were trying to get our arms around the developing shitstorm.

"We're working our liaison with the Task Force; not much yet," Alec spooled off as if reading from hastily scrawled notes. In the back of both our minds, we were kicking ourselves for not being able to get into the planning for Operation Anaconda, a significant Special Operations op to clear bad guys out of the Shahi-Kot valley

"From what we're picking up, one of the helos took heavy fire on landing—guns, RPGs. They got back in the air but couldn't hold altitude."

"Casualties?" I asked, fearing the worst.

"Unconfirmed. But it sounds like somebody fell out when they got hit." More paper shuffled in the background.

"Thanks," I said curtly. "Keep me posted."

I set down the phone and passed the info to the double-wide team, which immediately passed the info to Darran and the team in the GCS.

"We may have a man down at the initial point of attack, and we're probably the best asset to find him. We're working the CAOC to get in there, but in the meantime, keep looking around for the needle in a haystack."

I glanced at the clock. A bare few minutes had ticked by, minutes that felt like an eternity waiting for the gears of authority to give a grudging turn. It was hard to imagine what time felt like on the snow-covered ridge below. Even from this distance, we could see that the whole damn ridgetop was rippling with enemy fire.

I churned through the few facts on hand. Razor Zero-Three now sat in a smoking heap in the valley. The Navy SEALs on board were already humping their way back up the mountain to rescue the man they'd lost—seven miles straight up in deep snow with shit-thin air, with a lot of bad guys in the way. A flash of pride cut through the storm of emotions inside me. "No man left behind" wasn't just a marketing slogan.

The second chopper, Razor Zero-Four, flew straight into the meat-grinder to drop its own team of SEALs, call sign Mako 30, on the ridge. By the sound of things, the Taliban had a lot more guys on the mountain than anybody expected. A continuous rain of heavy machine guns and RPGs were pushing the SEALs back off the mountaintop.

A Quick Reaction Force (QRF) was dispatched from Bagram Airfield: two more Chinooks loaded with Army Rangers. In the civilian world, cops in trouble call for SWAT, guys who roll in on a situation with no advance planning or knowledge and simply wrestle control by force just long enough to snatch our guys out of harm's way. A QRF is largely the same thing; we surge in, suppress violently, and extract.

While a part of me was dying to get into the fight, the bigger part hoped that the QRF would end this thing in the next few minutes. For anybody on the ridge bleeding right now, time was life.

I did a quick assessment. Genghis was in the pilot seat. A top-notch F-15E jockey, she was one of my best in a fight. Tall and athletic, Genghis had a boundless love for the outdoors. I was lucky to have her on the team.

Andy was running sensors in the right seat. A young staff sergeant, Andy had learned a lot from Gunny, for better and for worse. He picked up bits of Gunny's sandpaper charm, but also Gunny's lethal effectiveness. Both Gunny and Andy had distinguished themselves as able to thread missile shots in through the window of a target building.

Darran was behind and between the two, informally parked on a cheap folding chair. He would be juggling maps and nav charts, working radios in a crunch, and doing anything else that couldn't be squeezed into the heads-down displays at knee level or the larger heads-up display screens at eye level.

Cliffy wasn't in the room, but his MacGyver magic was unmistakable. The normal GCS design brought no allowance for the display of threat data and imagery direct to the pilots and sensor operators. Cliffy had wrangled a way to split the threat data streams off, through a repeater, and into the GCS. The factory-original design demanded that threat data—we're talking in the middle of a fight—be printed off in the trailer and run out to the GCS on paper. Thanks to Cliffy, our pilots could see the fight play out in real time. It was another case of a million-dollar requirement stitched together with zip ties and Velcro.

Big was parked in the second GCS, watching the same feeds that were playing in the live one. That gave us a second pilot to swap when needed and allowed him to be up to the moment on situational awareness.

It was a fluke we were out there in the first place. Task Force Mountain, a 10th Mountain Division team, had been dispatched as part of a bigger effort to clear the remaining Taliban fighters out of the Shahi-Kot valley. It was all hush-hush, but based on the efforts to keep everybody else out of the airspace, somebody was planning a big fight. We were high overhead, transitioning from one collection target to the next, and all this just happened to be on the way. Basically, we rubbernecked into one of the biggest fights in Afghanistan.

On his own initiative, Andy began to scour the bleak mountains for any signs of life. All our sensor operators were great at sniffing out significant activity. He had been searching through the mountains when he came across an inbound Chinook helicopter and decided to follow it. "Hello," Andy muttered. "Where are you going?"

Andy glanced back at Darran and raised an eyebrow; Darran nodded back. The exchange was just another example of the odd-ball Vulcan mind meld we had developed working together.

Andy and Chris B. focused on the heads-up display with the MTS video, watching as the chopper rolled into a tight search pattern. Just a moment later, it slowed to a hover above a tall peak. Whatever it was looking for, it seemed to have found it.

Something found it as well. A flash of heat cut across the screen, a streak of white against a background of gray in the monochrome eye of thermal imaging.

"What the fuck was that?" Darran was in the mission-commander seat, leaning forward to stare at the screen as the glowing blob of helicopter thumped down hard on the mountaintop.

"Holy shit," Genghis snarled, "I think we just lost a chopper."

Andy zoomed in and out, slewing the camera to find the shooter.

A Chinook looks like an Oscar Meyer Wienermobile with rotors above both ends. As targets go, a hovering Chinook is like the broad side of a barn. The huge ramp door at the ass end of the chopper dropped open, and guys scrambled out, guns firing.

In the ghost-gray of infrared, we saw the explosion of black as the RPG slammed into the Chinook. Through the distant IR view, every rock, shrub, and crevice lit up with the starburst of muzzle flash. The helo lurched, side-slipped, and dropped out of the sky. It slammed down on the snowy peak in a spray of burning debris, the giant rotor blades slowly churning to a stop.

"Probably Razor Zero-One or Zero-Two." The intercom voice came from the double-wide.

Call signs were a best guess for us at this point, just what we could pick up on radio. We were twelve miles away. Our mission was farther north, our presence here little more than a drive-by. Being there when the shit hit the fan was only happenstance.

That fan was Takur Ghar, a rugged ten-thousand-foot peak jutting up along the eastern edge of the Shahi-Kot valley. This was Taliban country, the kind of rugged stronghold that helped predecessors stand off the might of the Soviet Union years before. Judging from the AWACS (Airborne Warning and Control System) overhead, and the AC-130 that had lapped through just beforehand, whatever was playing out on that jagged spike of rock was a big deal, something to which we hadn't been invited.

In an instant, Genghis was on it, immediately requesting AWACS clear us into the airspace. It wasn't budging. "They said they already had assets in the area," she called out. As best she could, she explained that we weren't some run-of-the mill observation platform—we had fangs. But despite her pushing, she was told to "stand by."

This was no SA-3 engagement, no creeping forward a slice at a time like we had done in the past. Respecting US airspace limits wasn't optional; anything sticking its nose uninvited into the kill box would get that nose shot off, with no care who we were or what kind of good intentions we brought to the fight. Genghis was forcefully pitching our case with the AWACS even higher overhead.

What might look to the civilian eye like a commercial airliner with a giant rotating pizza bolted to the top, an AWACS was a command and control aircraft, the all-seeing, all-knowing eye that provided air traffic control across the battle space. It had the final word on what flew through the giant box of sky beneath.

To say that we were not invited was an understatement. We had been emphatically ordered to stay the hell away. The airspace had been declared off-limits, and we respected that boundary. But like any good warfighter we hugged the edge of our assigned airspace and peered intently inside to see what was going on as we flew to our next target.

Time froze for just a second, barely a heartbeat, a moment that felt like it stretched out for minutes. Then reality snapped back with a face-slap of adrenaline and Darran punched up the LNO.

"Ty, grab Eric and Alec, we need to work this." Ty and Eric were going through changeover at this point. He didn't have time for conjunctions, and Ty didn't need them. Keys clattered like a machine gun as he worked with the Predator LNO at the CAOC in parallel.

It was time for me to swap seats with Darran. Alec hit the line first, reading my mind. Watching the same feed we had, he led off with, "Go direct with the ground; you've got whatever support you need from here."

That was exactly what I wanted to hear. With warfighters in mortal peril this was no time to discuss interagency agendas or conflicting mission priorities. On his own authority Alec had just chucked our existing task and committed the full weight of the CIA behind saving American lives.

"We're reaching out to the CAOC Predator LNO now," I responded, snapping hand gestures to people around the GCS. No matter what the urgency, barging into controlled airspace in a crisis was a sure way to get shot out of the sky. It's one thing to play cat and mouse with MiG fighters, but I wasn't about to have a midair with one of ours. I shifted my focus.

The guys on the mountain continued to take fire. *I don't think the bad guys were all that surprised,* I grumbled inwardly. By the looks of things, the bad guys had a pretty good idea we were coming.

Even from this distance we could see damage to both sides of the wrecked Chinook as rounds tore in one side and out the other. The inside of the helo had to feel like a garbage disposal. Bodies scrambled out the wide rear door, guns blazing. Some fought their way to scant cover; others dropped motionless on the edge of the ramp.

Still wrangling for clearance to enter the fight, we could only watch in horror.

Somebody on the ground started calling for air support. Identifying himself as Slick Zero-One, the beleaguered Airman had to split his bandwidth between moving, shooting, ducking, and coordinating an ad hoc air strike on what amounted to his own position. From ground level, he described the primary source of enemy fire as coming from a bunker some fifty to seventy-five meters uphill from the nose of the helo. Stealing glimpses between bullet strikes, he had to see the rise as a gravel wall speckled with muzzle-flash.

Our perspective from on high gave us the ability to provide a more analytic assessment. Absent much of anything for scale we reverted to our "unit of measurement" trick. The Chinook was a known, the fuselage measuring about sixty feet from tip to tail. The front face of the bunker was just over two Chinooks away, a distance closer to forty meters. We could see the irregular line of fighters fanning to either side of the primary bunker, tucked behind rocks or the sparse vegetation.

Almost every US asset in the sky was converging on Takur Ghar. Slick Zero-One engaged a pair of F-15E Strike Eagles that came at a run. I looked at the scene, knowing up front that talk-on from the ground would be an utter bitch. The Chinook was the only real point of reference, and the Eagle pilot would have to engage hostiles that hid within pistol-range of our survivors. The blast radius for the smallest bomb on the Strike Eagle would encircle the entire fight, friend and foe alike.

I pushed thoughts of Donnie from my mind and focused on the problem in front of me. Clearly on the same wavelength, Slick Zero-One was having a bad enough day without rolling the dice on that sort of crapshoot. He opted for gun runs.

Mind you, the guns on an F-15E are no joke. A dedicated killing machine, the Strike Eagle packed a General Electric M61 Vulcan Cannon. The modern progeny of a multibarrel Gatling gun, the Vulcan could spin up and vomit 20 mm explosive-incendiary shells at nearly a hundred rounds per second. In what might sound like a chainsaw revving, a Vulcan could saturate an area with bullets, each one bigger than a man's thumb. What wasn't shredded on impact was blown up and set on fire. The Air Force doesn't fuck around in a gunfight.

It took several minutes to get the Strike Eagle oriented, according to both where to shoot and where emphatically not to shoot. Absent a laser, almost everything relied on Slick Zero-One's ability to speak under heavy fire, the Strike Eagle pilot having only a glimpse of the target as he sizzled by at jet-fighter speeds. The Eagle rolled in. I could imagine the Gatling gun chewing a line of earth and rock into a trench of sand and gravel. Radio traffic from the ground reported some exposed enemies had been hit or scattered.

But the bunker held, the fire from within undiminished. The Eagle made a second pass, a third, a half-dozen or more with no change. Out of ammo and getting low on fuel, the Eagle had no choice but to break off.

The door of the GCS swung open abruptly. Marcella made a beeline behind Darran and Chris, then dove between Genghis and Andy. Without a word she punched a button, ejecting a tape cartridge with a short whine. Shoving a new tape in place, she turned on her heels and left just as abruptly. The door slammed behind her.

I looked at Darran, confusion on my face. He shrugged back, by all appearance equally baffled. Darran continued to work the Predator LNO to get us into the airspace. By the time Big swapped into the pilot seat, we received clearance into the airspace. He immediately began attempting to contact the ground party, which was difficult because we only had unencrypted radios and they were using encrypted radios.

Finally! I thought. *Now maybe we can do some good.*

Not much had improved on Takur Ghar. Although the F-15E's guns had helped to shave the edges off the enemy force, the core remained terribly effective. Even with the best of tactics and training, the crew of Razor Zero-One had only the ammo they carried, firing guns that seemed anemic against the 12 mm Soviet machine guns that pounded their position.

With no relief in sight, Slick Zero-One opted to swing for the fence. An Air Force F-16 Falcon was next in the batter's box, which like its beefier brother carried a Vulcan cannon. But Slick Zero-One had moved past the notion of winning this fight with a series of base hits; he opted for a Louisville slugger in the form of a five-hundred-pound JDAM, the same weapon that cratered Kandahar's runways at the start of the war.

No question, spiking a JDAM through the roof of the bunker would obliterate everything inside. The problem was that the blast would also shred the Chinook and everyone hunkered around it. Ending a fight is important, but surviving the ending is the whole point.

The answer, as best as they could reason, was to drop the bombs on the far side of the bunker. Placed properly, the blast would wash over the back of the bunker but come up just short of the Americans.

I thought about Alec's "Black Widow" and her bug splats, about all the science that goes into predicting where fragments of white-hot steel might go. That game doubled in complexity because we had no idea what else was inside the bunker. Secondary explosions, the kind we might get from, say, a stack of mortar shells, could turn a big explosion into a helluva big explosion. The margin of error was paper thin, maybe nonexistent.

I doubted that Slick Zero-One had the bandwidth for much of the math, but by the sound of things he had a real solid grasp on the concept. He directed the F-16 to drop on the back side of the peak, hoping, I supposed, to send enough of a shockwave through the earth to collapse the fortification. That strategy would put as much of the mountain as he could wrangle between his men and the blast. It was a ballsy move, but he was low on options.

The F-16 rolled in. Much like the Strike Eagle before it, the F-16 delivered its rounds spot on target. We watched the attacks that played out like inverted fireworks, a streak from the sky that burst explosively when it reached the ground. Huge chunks of rock were blown into the air, hot earth vaporized into a thermal cloud. Then the dust settled, and from within the bunker, the gunfire resumed.

A second bomb was dropped, the target point edging closer. Then a third, closer yet. The mud-spatter of rocks and debris rained down across the Chinook.

It was agonizing to listen to run after run, kicking up a lot of dirt. Slick Zero-One was getting punchy trying to call in fires and fight at the same time, having to repeat the same instructions to each new aircraft. He was running on the end of endurance while we were rotating bodies as needed. On a rapid rotation, Troy had just cycled into the pilot seat.

"I've got 'em!" The voice erupted from behind me, just a heart-beat ahead of the rush of fresh night air that flowed through the open door.

I turned to see Marcella once more in the doorway, now holding a piece of paper at head height. A fire burned in her eyes as she stepped in and slapped the page on the desk. "I know where they are."

I looked down at the page beneath her hand. It was a photo-graph, a screengrab from Predator's camera feed she had pulled off the tape. The image was overlaid with hand-drawn notations. Lines of sight, angles. And a red circle on the front face of a pile of rocks. Marcella stabbed a finger in the center of it. "We need to shoot here."

I winced, tapping a cluster of blobs on the photo. "No, Marcella. Our guys are here, and the bunker is—"

"I know where the fucking bunker is. But they're shooting at the wrong place." She shook her head, forcing her calm. "No, not wrong, just . . . they can't see things like we can. Look, those F-15s and F-16s have been hitting the same spot with everything they've got. Pounding the roof, pounding the back side—it won't work. The walls and roof are feet, hell maybe meters, of rock and earth. It's Taco Bell ten times over. We need to put a shot through the front window."

I looked up at her and blinked. A well-earned rule in warfare is that the guy on the ground, typically the guy getting shot at, has the best understanding of the battle space. As a nation we'd seen time and time again what happened when some REMF,[75] far

---

75 Rear-echelon mother fucker, a derisive term used by combat troops in a number of contexts, in this case when speaking about politicians and bureaucrats who try to stick their nose into an active combat scenario.

removed from the battlefield, thinks he or she has a better idea of what is going on. That happened in places like Mogadishu, and we all know how that turned out.

And yet here was an imagery analyst who could fit in a large rucksack telling me that from eight thousand miles away she had a more finite understanding of the battle space than the SEALs and Rangers who were bleeding all over it.

And fuck if I didn't believe her. Marcella wasn't a showboat; this wasn't some midcrisis play for relevance. I'd stand her up against the entire flock of NIMA-nerds that Alec had at his disposal. I'd seen Marcella identify a truck by its shadow from four miles in the sky. She raised an eyebrow and tapped the circle. "Right here."

I glanced at the control station. Next to Big at pilot was Joker in the center seat, Andy at SO, and Chris as the backup sensor operator. I pointed her to them. "Brief 'em."

I turned to Joker. "Get Slick Zero-One on the line."

What followed had to be one of the strangest combat discussions in history. I had no idea how Slick Zero-One would handle any of it. Here he was, stuck on a mountain in the middle of a helo crash, bullets chipping rocks around his head, and we're poking him to juggle off secure comms to answer an open line just to talk to us. I felt like a telemarketer from the Twilight Zone, calling the poor guy at the worst moment in his life.

Hi sir, you don't know me, in fact on paper we don't exist. But you're in a horrible firefight right now, and while technically we are eight thousand miles away, we have an unmanned airplane over your head packing a pair of antitank missiles. What's that? Uh, no sir, we don't actually use the term "model airplane."

Once we got past our identity, we explained why the gun runs and bombs had failed and how a Hellfire through the front door would do the job. Remarkably, Slick Zero-One had some knowledge of Predator as a platform, but the Hellfire part was a twist. What we proposed was not going to happen at the far edge of the envelope; it was gonna happen just off the end of his nose.

Now, skepticism is healthy. It is critical in the military and never more so than when you bring explosives to a firefight. Nothing is friendly about friendly fire, and nobody in his or her right mind is comfy with the idea of calling for high explosives to be delivered less than a football throw away.

But if necessity is the mother of invention, getting shot at is the mother of opening your mind to alternatives. Having burned through four manned fighter's worth of air support to no avail, the relentless barrage made any idea, even a crazy one, seem a little more palatable. Still, even neck-deep in shit, there are limits to how far somebody is willing to make a leap of faith.

Joker cupped his hand over the boom mic on his headset. "Slick wants to take the shot, but his ground commander wants a demo shot on some bush down the hill."

I scowled. One round of show-and-tell would expend half of the firepower we were carrying. "He understands we only have two missiles, right?"

Joker nodded. "I made that real clear. Slick's onboard, but the GC says it's the bush or nothing."

I cursed under my breath, hating to waste fifty percent of my ordnance on a demo. Still, I couldn't fault the guys on the ground. Stuck between a rock and a hard place, they were trying their damnedest to save lives. If I were in his shoes, I'd have done the same thing.

Sometimes, while in a fight, we need to make a call quickly, without the normal string of "Mother, may I" permissions. This was one of those moments. I turned to Joker. "Hell, if that's what he wants, shoot the fucking bush."

Everyone spun into action. Our load-out carried the usual pairing of two missiles, a Kilo and a Mike. The former did a better job of punching through armor, while the latter would throw a hell of a lot more violence across an open area.

"Use the Mike on the bush," I said to Big. With the new sleeve on the K, it was my biggest hitter. We could show off with the jab and keep our roundhouse for when it counted. If we did fire on the bunker, I wanted to turn the inside of that thing into a blast furnace.

Andy swung the optics to a lone scrub bush that clung to the barren slope a couple football fields away. As the crosshairs centered on the straggly plant I asked him "Are you sure we're all looking at the same bush?"

"Fuck if I know," Andy replied, pushing the zoom. "It looks like the most prominent one I can see from up here. But I've never had a talk-on to a plant before."

Fair point, I thought. Low scrub was largely IR neutral, pretty much the same temperature as the rest of the environment. For all we knew, we could be aiming at a completely different bush. But there was no time for dragging this out. We ran the shot by the numbers, the bush was a target like any other until the sudden stutter in the video feed as Troy squeezed the trigger and the missile left the rail.

Then came the wait, twenty to thirty seconds as the missile streaked down out of the sky, waiting to see if with millions of dollars we'd been able to build a transnational laser-guided Weed eater. With nothing really to lock on to, Andy had to hand fly the missile into the bush.

The blast of impact was anemic in IR, not much more than a dull flash that scattered a lot of hot gravel. There were no car parts, no fuel tanks, no burning debris. But as the dust cleared, we could see the burning scraps of an Afghan tumbleweed scattered around the impact crater.

Moments later, Joker looked back at me, flashing a thumbs-up. "He wants us to hit the bunker!"

"Great!" I said, more in relief than irritation. It was excruciating to watch Americans under fire, feeling we could help if only given the chance. Only now we weren't just a chance; we were pretty much the last-ditch hope.

I got Alec back on the line. Making the call on a bush was one thing, but unilaterally slinging a missile right over friendly troops was a different matter altogether. I needed somebody higher up the food chain to authorize a shot that had a chance to end this fight.

Alec was quick to relay the decision. "You've been cleared to engage." He had no need for a "don't fuck it up" here. He understood just how paper-thin we were slicing it.

"All right, here's our target," I said, handing Big the sheet with Marcella's marked-up photo. The talk-on we were getting from Slick Zero-One was accurate, just not specific. "*The bunker fifty feet in front of us*" set up the ballpark, but the gap under the big slab of rock was the sweet spot, the Achilles' heel. All we had to do was hit it.

There was no point trying to explain it to the guys on the ground. The long and the short of it was that we were about to send a Hellfire missile screaming over their heads. Trying to explain it to Slick Zero-One would have been like trying to describe a one-mile sniper shot to a guy in a nose-to-nose pistol battle.

In this case, though, the sniper analogy was a little unsettling. We weren't a mile away; we were over two miles above a ten-thousand-foot peak. With only fifty meters separating a hit dead-on the enemy from one dead-on our own men, our margin of error was somewhere on the order of one-quarter of one degree. Factor in for the blast and frag radius, and we could cut that margin to about an eighth of one degree. In sniper parlance, a one-eighth minute of angle shot was akin to hitting an aspirin dead-center from a football field away. Miss that shot by just the diameter of the bullet, and good guys die. Add to that freezing cold and high-altitude winds, and a team that had been on-station for the last fifteen hours. But if we don't take the shot, everybody dies. Nah, we had no pressure at all.

But if anyone around me was aware of the difficulty, it didn't show. I watched my team perform like it had on every other shot, with skill and precision and no skipped steps and no shortcuts. Nobody was unprepared. It all came down to this.

The optic ball tracked smoothly as Wildfire swept in, and I ran through the final checklist: "Missile Armed. Laser on. Lasing. Confirm Lasing."

Gun flashes burned from inside the bunker, the massive forked flare of a heavy machine gun centered on the screen.

American voices were on the radio, nearly drowned out by the chatter of gunfire.

"3 . . . 2 . . . 1 . . . Rifle."

Big pulled the trigger and the missile vanished, nosing up only briefly before seeking out the shimmering ball of laser light that Andy held on the target. Time slowed to a crawl, my entire being focused on the image on the screen. One-eighth of one degree.

Then a black flash exploded, debris and body parts flying through the air. Dust and smoke obscured everything.

From within the haze I heard screaming; raw, ragged voices on the radio. I couldn't process the words at first, only that they were English—American voices. It took a long moment before my brain latched on to an explosive "Fuck yeah!" Not screaming, I realized, but cheering.

Only then did I remember to breathe. The image on-screen cleared as mountain winds swept away the cloud of dust. The men of Razor Zero-One were standing up from behind their spartan cover, guns shouldered and sweeping the bunker. But the only thing coming out was smoke, trails of black and the occasional spark of loose ammo cooking off. Whatever had been inside was dead.

A roar tore through the GCS, the same torn-from-the-gut cheer that likely was rattling the walls of every structure involved with this process. Dots seemingly scattered randomly across the globe all tasted victory.

It was a short celebration. We'd gutted the Taliban fortress, but a whole lot of bad guys were still clambering around on the mountain. With a much-needed breather, Razor Zero-One had a chance to regroup and address the wounds. The soldiers were low on ammo and battery power. We maintained contact as their two medics worked frantically to stabilize the injured; from the sound of things a couple guys were pretty bad off. Slick Zero-One continued to call in air strikes against the clumps of enemy forces scattered across the mountain.

We had no word on incoming helos and wondered where the evac team was. Joker stayed on the radio with Slick Zero-One, a friendly voice to lend some comfort as the time dragged on.

"We're freezing up here," Slick Zero-One said quietly at one point, "and another guy just died. Can you talk to the Army and get us off this mountain?"

Although we were out of missiles, we made the call to stay on-station, circling overhead to provide the best set of eyes. I heard Joker say, "Slick Zero-One, do you need me to take over? We'll handle it from up here. We're not leaving you."

Slick Zero-One seemed all too relieved to have the help. It was pretty apparent that whatever the hell this Wildfire thing was overhead, we had earned our chops with the guys on the ground and with the AWACS as well, it seemed, who were all too happy when we started calling in other air assets to drive the enemy off the rest of the mountain.

An al-Qaeda convoy tried to push its way onto the ridge; it died ugly. Joker had been lining up a pair of F-15Es to do a gun run on the convoy when a Spectre gunship declared itself available. Troy kept our Predator in a perfect ellipse over the battle space while Joker called in fire. Using our Rover feed as a guide, the AC-130 fired what looked like a laser beam of tracer fire, carving trenches in the earth. Running or hiding, if something escaped one round of carnage, Andy picked it up and the cycle would repeat. It was less a matter of clearing the mountain as it was sanitizing it with fire. As I watched the onslaught I found myself doubting that so much as a scorpion could have survived.

"Where the hell are the evac birds?" The question became a mantra in the GCS as the hours slid by. Razor Zero-One was relentlessly demanding medevac, declaring status critical for at least one of the wounded. But with Bagram just over an hour away, we still weren't tracking inbound choppers.

It came as no surprise that by now, somebody well up the chain of command would be damned curious about their wayward bird and why it wasn't headed home. For the umpteenth time, I had Alec on the line, echoing Joker's commitment to the guys on the ground.

"There's still no word on the evac. Tell everybody whatever you need to, but we're not leaving 'em alone up there."

Alec didn't need convincing, he was shoulder to shoulder with all of us that this mission trumped anything else going on. Fuck the priorities; today we are saving a handful of guys stuck on a mountaintop in the ass end of nowhere. We had unmatched situational awareness and the ability to designate fire on just about anything that moved. If anybody with ill intent even thought about coming up that slope, they'd only live briefly enough to regret it.

I told the crew, "I want iron on the mountain. Let enemies and the friendlies both know that airpower is here." And by God we rained iron.

The barrage fell into a rhythm with the steady stream of air assets cycling into the battle space. Joker had planes stacked like pancakes: Strike Eagles and Falcons, Navy F-14 Bombcats, Marine F/A-18s. While Andy was buddy-lasing one target, Joker established comms with the next inbound, exchanging the laser codes necessary to allow one aircraft and bomb to tune to the laser illuminating the intended target. This allows simultaneous or nearly simultaneous attacks on multiple targets by a single aircraft, or flights of aircraft, dropping laser guided weapons while limiting spoofing by the enemy. That's a technical way of saying that we became the express lane for the delivery of American firepower.

To stay at our best, we hot-swapped pilots and SOs, keeping fresh eyes on screens and steady hands on the controls. Joker stayed in place at the center seat, having established himself as an authority presence in the battle space. He had talked so many aircraft into the fight that he had it down to a science. Perhaps as important, he had become a persistent point of contact with the guys on the ground. "We're not leaving you guys," he repeated time and again.

We did all this largely on our own authority, side-stepping the CAOC and the interminable delays of asking for permission at every turn. Nobody argued, nobody bitched; it was a group of Americans doing whatever it took, big or small. I couldn't imagine what sort of voodoo Alec was doing to placate his side of the planet, but everybody backed our play and let us do the right thing, the American thing.

While our ability to remain on station seemed limitless compared to that of our gas-guzzling, jet-powered brethren, our promise to stay to the end was finally challenged by a glowing Low Fuel indicator. Big, now in the pilot seat, turned to me and sadly announced that we had entered the last stage in the evolving path towards bingo fuel. The time had come for Wildfire 54 to head home.

Had this been a manned aircraft, this would be the end of our part in the story. Out of gas was an unforgiving and unbending rule that existed since the dawn of powered flight. But Wildfire had made a promise to these guys, and we didn't have to play by the rules.

For the last several hours, our second Predator, Wildfire 55, had been winging its way into the battle space. On what amounted to a "3, 2, 1, go," we executed a Chinese fire drill between the two GCSs, pilots, and sensor operators swapping seats with their counterparts. In just moments, the crew familiar with the fight found themselves at the control of a fresh aircraft with full gas tanks. Like I told Alec, we weren't leaving these guys, even if we had to jump our crew from one fucking plane to another in midflight to stay on station.

And it was good that we remained. Despite everything we had thrown at the enemy thus far, they remained a tenacious opponent. A second bunker to the south revealed itself and suddenly started lobbing mortar rounds across the ridgetop.

Mortar attacks are a cause for urgency. With the ability to rapidly arc one shell after another, a crack mortar team can zero in on a target pretty damn fast. We weren't about to give them the time. Reaching for the top card in our deck of available assets, we blinked with surprise when it came up a French Mirage 2000.

That's not a knock on our coalition brothers; the French flew with distinction. But although this process had become almost business as usual with US air assets, handling a talk-on and potential buddy lase—sharing the laser from one asset to guide the weapon from another asset—with a foreign national in this case, even a trusted ally, was a different matter. The French pilot wanted our laser code; we replied that we'd adjust our laser to match his bomb. It was a short battle, and we weren't budging. Suffice it to say that the French crew proudly added their fist to the beating that fell on the southern bunker.

The knock-out blow came from yet another Navy Tomcat, the swing-wing fighter made famous in the 1980s film *Top Gun*. Since the bunker was located considerably farther away from our guys on the ground, we were free to go heavy on the ordnance. The mortar team was loading another round when a one-thousand-pound Paveway II turned the entire pit into a giant steaming golf divot. As chunks of debris and mortarman rained from the sky, I felt a bit of old-school satisfaction that touched me down to my Air Force blue.

For the impossible shot, use precision. For everything else, drop a shit-ton of high explosives.

As promised, we remained on station into the night. When nothing was left to shoot, we were simply a friendly voice in the darkness, a presence overhead. Scott and Gunny were now in the seat, with Will, Steve H., and Rob G. swapping out. We did more rover work with an AC-130, scanning the helo ingress route for threats.

In the spirit of "use every blade in the pocketknife," Gunny came up with an innovative angle for yet another part of the Predator package. The MTS ball had not only a laser designator, which we had been using to great effect all night, but also a laser illuminator. Invisible to the naked eyes of the enemy, US troops with night-vision saw the illuminator as a powerful spotlight that shone down out of the night sky.

Gunny repeatedly swept the peak, pushing back the darkness. It was less about discovering a threat than it was giving the guys on the ground that extra measure of comfort knowing we were overhead, still watching. In the end, the IR spotlight helped guide in the rescue choppers that at long last arrived to take our guys home. The pickup and exfiltration were uneventful save for the result: Slick Zero-One and his buddies made it off that godforsaken mountain.

At some point, Colonel Boyle asked if the sun had risen in Afghanistan. After untold hours in a darkened GCS staring at infrared camera footage, the question took everyone by surprise. Gunny flipped the camera ball to optical mode and the stark reality of the "morning after" hit home.

Our flight team was exhausted, but we could only imagine what the guys on the mountain had endured—not only the long hours but also getting shot at, crashing a helo, and enduring the bitter cold and loss of brothers. Our exhaustion didn't seem to stack up.

In the course of what had thus far been the sum of the Predator program, contact with friendlies on the ground had been rare by design. In the jargon of the CIA, "We were never there." Any awareness at all was more often than not the nameless provision of a laser dot before we disappeared into the night.

But this fight had been different, the unexpected but unforgettable payoff coming in just a few words of radio traffic between Slick Zero-One and the air assets overhead. When asked a question about the status, Slick Zero-One replied, "Talk to that Wildfire guy. I don't know who he is, but he's been saving our ass all night."

16: OBJECTS IN THE REARVIEW MIRROR

ALEC BIERBAUER

WE FEW, WE BAND OF BROTHERS
September 2010

I walked the length of the second floor of the Smithsonian National Air and Space Museum, taking in the history of man's eternal quest to fly. The Bell X-1 sat poised in all its blaze-orange glory. In 1946 a young Chuck Yeager strapped his ass into that thing and became the first man to blast through the sound barrier. At the time, nobody was sure if the plane would keep going or just explode. But just some twenty years later, Pete Knight clocked almost seven times that speed in the deadly looking black X-15, hanging just over there. To my knowledge that record has never been broken.

Just about everything in this enormous building was a first. The *Wright Flyer* was down on the first floor; it was hard to imagine that spindly assembly of wood spars and cloth shared any heritage with the jets overhead. The *Spirit of St. Louis* was here, the first plane to make a solo crossing of the Atlantic.

My eyes fell on the tail of the *Spirit*, where the designation N-X-211 was stenciled. I laughed, the all-but-forgotten memory returning. It was near the end of our program, when Predator began to expand from its Afghan debut and spread its wings over the Global War on Terror. Predator employment, in many capacities, would spread across the world, to other theaters and other missions.

On one of those early missions, we became the focus of a zealous air traffic controller who demanded to know who we were and why the hell we were in his airspace. Mark talked to him, a protracted dance of mostly side steps on our part, until the ATC demanded our tail number—or else. Mark's sense of humor may have kicked in, or his confidence that he knew more about flight history than some angry guy in a distant tower, so Mark gave him N-X-211. Apparently satisfied in his victory, the guy jotted down the number, and we were once again on our way.

A familiar voice cut through my reverie. "Hey, Alec."

I looked up to see Mark's smiling face. He looked fit despite the passage of years, thus far staving off the few extra pounds or touches of gray that had inexplicably found their way among some of us. The reflexive handshake gave way to a big hug. He didn't chuck me over the railing, so I assumed that whatever abuse I might have thrown his direction years ago had been forgiven if not forgotten.

Mark looked to his left, and I followed the gaze to a group of men and women gathered around a very familiar sight.

"It's her," Mark said with a hint of reverence. "It's Thirty-Four."

I felt a funny rush, like unexpectedly bumping into my first girlfriend. "You're kidding me." I knew the museum had a Predator on display, at least a reasonable facsimile. I couldn't imagine that any of our birds had actually survived to see the light of a cushy retirement in air-conditioned comfort.

But there it was—not just any Predator but our Predator. The decals were new, the obligatory Air Force and Big Safari logos plastered over what back in the day had been its unblemished gray skin. But it was the one. My eyes took it in like a long-lost friend.

A wave of smiling faces closed on me, with more hugs and slaps on the back. Ken Johns and Snake, still side by side, showed up, as did Brian and Terry McLean. Joker was there, his fitness a given as he was still active duty—from what I heard, a rock star who continued to rise. I struggled to pick out faces I knew from the rest of the crowd, likely husbands and wives, sons and daughters, some probably seeing what we had done for the first time.

"How the hell are ya, brother?"

I turned at the familiar voice, knowing before I did that I'd find Cliffy just as I remembered him: T-shirt over jeans, wearing a big earnest grin. Cliffy got a bear hug.

Troy was there, dapper in sport coat and slacks, his appearance reflecting the precision that infused everything he did. We didn't know it then, but this would be the last time most of us would ever see him. He passed away in 2015 doing what he loved most, flying. His plane continued on autopilot for another four hundred miles or so before crashing in the Pacific. Knowing Troy, I guess he just didn't want the ride to end.

Gunny was gone as well. We'd all been shocked by the news of his untimely passing in 2005. That left a big hole in all of us.

"So, what do you think?" Standing shoulder-to-shoulder with me, Mark posed the question while looking at Thirty-Four.

"About what?" I replied. "The plane, the team . . . ?" It was a little overwhelming.

"All of it—what our team did." Mark seemed deep in thought, and for good reason. I doubted that anybody could have imagined the reality that would spiral out of our little science project and our hunt for a single man.

I'd gotten a first-hand taste of that impact in 2002. My time with Predator had ended, and I went back to Afghanistan to be the

CIA lead for mobile team of four SEAL Team members, a tactical SIGINT team, and another colleague from the Agency. Supported by a squad of hand-picked Afghan security guys and a half-dozen Hilux trucks, we were sent to play high-stakes hit and run with the Taliban in search of some high-value targets.

It was a stark lesson on how much less we see on the ground than flying over it. Every rock, every tree, every swell of terrain was an obstruction that blocked our lines of sight. With speed as our only ally, moving through a battle space where we'd be sure to be outnumbered in almost any encounter, our day-to-day reality was charging forward into places we couldn't see. If a surprise waited around any given corner, it was likely to be a bad one.

My knowledge of Predator ops went past the technical. I'd also brought its phone number. So when the call came to coordinate a hasty strike on a high-value target, I did my best E.T. and phoned home. When a group of over fifty SEALs and Army Rangers showed up to pick a fight, we had a giant eye overhead. Beyond anything I could have hoped for, the GRC mission manager was a former high-ranking Army Ranger, and a dear friend. He was able to read out in front of the coming fight better than anybody I knew. If an ambush was waiting for us in the darkness, the ambushers would pay a price.

Now admittedly, of the numerous ways I'd seen Predator help guys on the ground, providing turn-by-turn GPS directions had not yet made the list. So when our convoy of highly trained special operators made a wrong turn in the serpentine riverbed that we used to travel invisibly, the all-seeing eye came through.

"At the first possible wash, make a U-turn."

It wasn't our most shining moment, made worse because everyone in the GRC was watching. An unwritten code was that "what happened in Afghanistan stays in Afghanistan," but our blunder was captured in high-res, full-motion glory. I knew I'd hear about it and probably see it again—and again, and again.

After completing a "tactical one-eighty" we rolled in on the target compound. We had pieces of intel on the layout, thanks in great part to two utterly Western-looking idiots who, putting mission ahead of brains, had dressed up in Afghan manjamas. Armed only with pistols, they drove boldly around the enemy camp in broad daylight taking copious notes.

What followed was a fierce two-stage breach. The outer gate fell to our Ranger Humvees while the inner gate, which I had noted as having "unknown size and composition," was treated as a reinforced door and hit with high explosives. Standing way, way too close when the blast wave slammed into me, I was reminded to be more precise in my future estimations. But the gate was obliterated, and our alpha wolves rolled in. The HVT was inside and decided to stand his ground against a bevy of SEALs and Rangers. He chose poorly.

Predator was there through it all, watching our backs, watching the far side of the compound for squirters, maintaining a chess master's view of a fight that we saw over the muzzle of a gun. When the smoke cleared, I stood winded, ears-ringing, riding the ebbing crest of adrenaline And I couldn't imagine how soldiers ever went into battle without that support.

I snapped back to the present as a hand clapped me on the shoulder, and I turned to see Snake presenting me to some guy in a nice suit. Offering me his hand and a big smile, he introduced himself as the curator of the National Air and Space Museum. He led off with, "I had to come meet you guys" and went on to talk about how our team had changed the face of aerial warfare.

The change was profound. When I left the program, we were running our one and only Combat Air Patrol. As I stood in front of the display, that number had risen to over fifty, maybe sixty, all around the world. Predator had become a much-desired, if not altogether mission-critical, part of global operations.

My eyes tracked around the faces in a circle, gathered once again in the shadow of our plane, and I was proud—not for anything I had done, just proud as hell to be among them.

I doubt any one of us imagined the Predator's evolution over the last decade. Our bird went on to blaze a trail across the global war on terror that took it to some rough territory off the Horn of Africa. In places like that, challenges come at the most fundamental level, like struggling to carve out five thousand feet of decent runway.

But it also had brought some welcome highs. Linda was one of those, everyone's top pick for Security Officer of the Century. With the program from the very beginning, Linda marked her tenure with nothing short of an extended series of miracles. She protected our secrets with a Doberman's tenacity, often through hard work and now and then by knowing just what regulation could be interpreted in our favor. I've never seen anyone manipulate a security system, or game it to our advantage, better than Linda. Her wisdom and work ethic were a godsend, and we were lucky to have her.

Another real treat was the chance to meet an old friend, Slick Zero-One. Then the disembodied voice of a tough-as-nails staff sergeant fighting to save his guys on a mountaintop in Afghanistan, Major Gabe Brown continued to excel in the Air Force. Two kinds of guys are in the special-operations worlds. The first is the movie version, the bigger-than-life muscled-up tanks who emanate badassery. The other kind is the quiet, unassuming ones who strike you as a great guy to go fishing with. Gabe was solidly in the latter category; you couldn't help but like him. His valor on Robert's Ridge had been recognized with a Silver Star.

Mark had gone straight from the double-wide to the commander of Air Combat Command's 547th Intelligence Squadron to turn everything he had learned from our experience into codified Air Force doctrine to train the next generation of drone warriors. His experience and insights built onto the Air Force "intel bible," adding Remote Split Ops to its already formidable toolkit. That assignment gave him the good fortune to work once more with Cliffy and Paul Welch; I envied that.

Following the development efforts of a clandestine Predator platform, weaponization, the initial year of use in Afghanistan, and expansion into the broader global war on terror, it was time for me to get into the field and continue as a more traditional operations officer. My tours of service had marched on through a series of war-torn or undergoverned countries in Eastern Europe, Africa, the Middle East, and Asia. These were countries being exploited by our adversaries, including the only nation in the CIA World Factbook to have absolutely no natural resources whatsoever. As a bonus, it held the uncontested record as the hottest place on the planet to ever have a US embassy. Unless we open diplomatic relations with the ninth circle of hell, that record is likely to stand. Surprisingly, as miserable as the environment might have been, I found myself surrounded by gracious and hospitable people, as well as a bunch of French legionnaires.

My work there threw me back into the deep end of politics and negotiations, more trade craft than tactical. I found myself dealing with authorities running the gamut from the local Department of Interior to the French Foreign Legion Commanding General.

It may come as a surprise at this point that I went into the CIA and not the State Department for a reason. That reason came out in a big way when we were negotiating with a foreign government to be granted the use of an airfield, which to that point had been in the hands of the French. Though the other nation was sovereign, France never hesitated to very pointedly remind it of the assistance France provided, which hadn't won many points. I saw this as opportunity, but it would require a delicate touch. Luckily, subtlety was my middle name.

With the enthusiastic support of the US ambassador we made some . . . "adjustments" to the meeting room where we would discuss the matter with all parties. The State Department, and, more important, our program, was lucky to have such a great ambassador and his well-earned positive reputation. The French arrived to find the long conference room in an odd state of asymmetry. On the right side of the table, where the American flag flew, were high-backed wooden chairs with leather cushions jacked up to throne height. Glasses and pitchers of ice water sat at each position.

On the left side, where a French flag flew on a shorter staff, was a row of straight-backed, uncushioned chairs cranked down to the floor. Only bare wood could be seen on their side of the table. By the looks of it, one might expect the delegation from Munchkinland.

The crowning touch, and the one that nailed the point home, was that the seat at the head of the table, where the local government official sat, looked exactly like ours. Like I said, it was subtle, but it got us our airfield.

"Hey, Alec, move over to the rail, they want to take some pictures."

The words broke my reverie, and I looked up to see that our group had spread itself along the railing in front of Thirty-Four. I shuffled into a gap between Brian and Mark. Snake parked himself just ahead of Brian, holding his hand as if introducing the plane.

Mark bumped into me, whispering through his camera smile. "You know, we oughtta write a book about all this someday."

"Pffft," I blurted. "I've seen your writing. How about a coloring book?"

"Fuck you," Mark chuckled, shaking his head. I was glad to see that some things didn't change over time. Then he added, "I'm serious."

I thought about it for a moment. Actually, I'd thought about it off and on many times over the years. Things had come out about the program, a lot of second-hand facts cobbled together, bits and pieces of grandstanding by others. But nobody had told the story the way we had lived it. I wanted my kids, maybe my grandkids someday, to know what I'd done with my life.

"I can probably get Mike to write it." I said.

"Who?"

I tossed a nod at a guy in the crowd, snapping photos. "Black polo shirt, big camera. We just finished working a USSOCOM program together."

Mark followed my gaze. "What the hell do you know about working with a writer?"

"Not a damn thing, but how hard can it be? We tell him our story and give him the One Rule."

"What rule is that?"

I grinned. "Don't fuck it up."

Mark laughed, then gave a slow nod. "That works for me. Let's talk when I retire."

"Deal."

My focus returning to the lenses in front of us. I smiled for the cameras, holding the pose as family members tried to capture the moment from different angles. It struck me that we were so few, just a handful gathered here to represent so many who had given so much to make all this happen. I dearly wished that more of us could have made the reunion; many absent friends were so deeply missed. I thought of the team, and of our mission. At the end of it all Usama bin Laden did not die in the blaze of an exploding Hellfire. But I took consolation that, in the last moment of his life, when all he could see was the muzzle flash from a Navy SEAL who punched a bullet through his brain, a Predator was overhead.

AFTERWORD

LIEUTENANT COLONEL GABE BROWN

People speak of angels who watch over us in a metaphysical sense, but on March 4, 2002, an actual physical angel loitered in the sky above Takur Ghar, a snowy ten-thousand-foot mountain peak in Afghanistan's Shahi-Kot valley, flown by a dedicated crew of angels operating out of a shipping container and a double-wide trailer eight thousand miles away.

This crew, and this piece of emerging technology, became the saving grace for American Air Force, Army, and Navy Special Operations Forces pinned down on that snowy mountain peak in the fight of their lives, during a mission that has become known as the Battle of Robert's Ridge. I was the USAF combat controller on the ground directing dangerously close airstrikes that fateful day March 4, 2002, when seven great Americans made the ultimate sacrifice:

USAF TSgt John A. Chapman
USAF SrA Jason D. Cunningham
USN PO1 Neil C. Roberts
US Army Cpl Matthew A. Commons
US Army Sgt Bradley S. Crose
US Army SPC Marc A. Anderson
US Army Sgt Phillip Svitach
Lest we shall never forget!

As an American Air Force combat controller, I had heard rumors of an enigmatic unmanned aerial vehicle carrying Hellfire missiles, and I am forever grateful that the crew of Wildfire 54 happened by that day. Wildfire 54 assessed the battlefield from above, and this was truly a game changer and life saver for American troops on the ground. In addition to being an eye in the sky, this UAV had the ability to loiter overhead for hours and provide laser-guided munitions. This Predator drone and the crew operating it thousands of miles away became the instrument that brought the fight of our lives to an end with the successful extraction of twenty-six American Special Operations Forces and the recovery of seven great Americans who gave their last measure in the name of freedom on that snowy hilltop on March 4, 2002.

This book tells the remarkable story of the humans behind the Predator drone, a motley crew of great Americans who fought myriad technical and political challenges and ultimately were instrumental in creating a piece of engineering that would forever change the battlefield for the benefit of American troops on the ground. This is the firsthand account of how the Predator drone became the omnipotent eye in the sky, surveying and conducting precision airstrikes in the name of American freedom and security!

Lieutenant Colonel Gabe Brown
United States Air Force

POSTSCRIPT: THE LONG AND WINDING ROAD

Opportunity knocks when you least expect it, and it is rarely appreciated or understood at the time. In the almost twenty years between September 11, 2001, and this writing, countless stories have been told about the rise of the drone revolution. Like all good fishing stories, after many retellings, many have entered the realm of wild exaggeration. Other stories were assembled like jigsaw puzzles through a process of research and interviews, some of them well-intentioned efforts that have suffered from the vagaries of differing or limited perspectives. As acronyms like UAV and UAS (Unmanned Aircraft Systems) and terms like *drone* have become part of everyday language, events accomplished by few have been claimed by many. The old axiom "failure is an orphan, but success has a thousand fathers" holds great truth. This version is a firsthand account, told straight from the trenches.

Still, even those who lived through a given experience rarely enjoyed an omniscient grasp of the situation as events actually played out. Rather than sweep away the fog of war through the clarity of hindsight, we've chosen to tell this story as we lived it, presenting truth as we understood it at the time.

While told through two sets of eyes, this is the story of a much larger team of remarkable people in which we were so very fortunate to be a part. The words that fill these pages cannot begin to truly convey the tireless work, sacrifice, courage, and initiative brought to the fight by this group of patriots, any more than words could convey our pride to be counted among them.

One might ask why this story has been so long in coming. The authors and contributors to this document have remained silent while continuing to work behind the scenes for their common cause—the United States of America and the freedoms we enjoy.

Yes, a number of books, interviews, and speeches about the technologies, tactics, and policies of drone warfare have appeared. Those stories have been presented, and in some cases manipulated, by a vocal few. Woodward, Franks, Tenet, Wright, Schroen, Crumpton, and others have described event chains that ran through al-Qaeda, the 9/11 attacks, and Iraq, highlighting the decisions of Bush, Clinton, and others. In those writings, aspects of the Predator program have been disclosed and often hailed. But in many cases the details have strayed, due in part to the separation between the respective storytellers and the actual events. Until now, nobody has completely captured the actual events, milestones, enablers, and obstructions. The story has never been told from the inside, in this detail, by the people who lived it.

Filling that gap wasn't an easy decision. The war against terrorism remained an ongoing battle, and we were adamant in our refusal to disclose any information that might be of value to the adversary. But by early 2016, we thought the time had come.

Technical details mentioned in this book were already in the public domain, or have no consequence on national security and the safety of America's war fighters. Besides, the meat of a great story is not in the nuts and bolts. We chose to give you a firsthand look into the reality of living through this remarkable effort: the sweat and tears, the setbacks, and the incredible victories through the eyes of the folks who made it happen.

If a core theme persists, it is that for the most part, senior leaders were not the ones driving the program. The lion's share of that credit goes to the midlevel men and women who all too often set aside career ambitions, deviating from the status quo of a safe progression through the ranks, to be a part of what would become a revolution in war fighting. Most are what we know to be rare people—individuals motivated by their patriotism, sense of duty, and desire to fight political correctness and organizational stovepipes as much as to fight the enemy. The stellar collaboration was between members of no less than fourteen different organizations: the US Air Force, CIA, DIA, NSA, US Army, USSOCOM, USCENTCOM, JCS, OSD, US Navy, NIMA (now NGA), NSC, US Marine Corps, and NRO all pulling for a common cause. Some of those organizations had multiple subelements involved.

The developments highlighted in this book are the product of science, intelligence, and determination. The subsequent evolution of unmanned weapons of war was so far beyond the immediate objective as to be largely irrelevant at the time. Use of these tools to prosecute lethal action against terrorists across the globe is, and will remain, a controversial issue. Although we set out to do so against one specific individual, we take great pride in being the impetus for the tools that have evolved to protect our service members and intelligence officers who go into harm's way to defeat terrorists in their safe havens. The due-diligence process, and the burden of how and when to use these tools, is and should remain immense.

These changes have garnered both adamant supporters and vehement critics. History shows us that revolutionary improvements to the tools of war are rarely characterized as "fair play" by an adversary. From the days of British longbow archers sniping French knights from afar to the advent of stealth aircraft that ghost through the sky, there has been a natural emphasis on minimizing risk to one's troops while maximizing lethal effect on the enemy. We stretched that line to the far side of the planet.

Sadly, our adversaries are equally adept at changing their tactics in response to our actions. Flying airliners into buildings was an out-of-the-box idea that caught us off guard. We should never again be complacent, nor should we ever stop evolving our ability to defend our great nation and our way of life.

Our professionalism demanded that we secure approval from anyone we sought to mention in the book. Some opted out and have gone without recognition, no matter how well deserved. Many of the patriots who do appear were referenced either by first name only, first name and last initial, or what may come across as an eclectic mix of military call signs and pseudonyms. We understand that this may have made the story a little harder to follow at times, but the decision was made with respect for the security of those still serving our country and for the safety of their families.

We are excited to know that people who were essential to our successes when holding GS-9, E-5, and O-3/4 levels in 2000 are now at the level of GS-15, SIS, chief master sergeant, colonel, or general. We watch every day for their next revolutionary, game-changing accomplishments.

Whereas the nature of a drone program is to protect Americans from loss of life and limb, we have suffered losses of our own since the events of this book. During 2002 operations in a classified location, an entire US Marine Corps KC-130 flight crew was lost in the process of bringing us a vital supply of fuel. They crashed in the mountains on approach to the airfield.

We have lost other members of our little family over the years, like Master Sergeant Jeff "Gunny" Guay, Troy Johnson, Mike Johnson, Brad Clark, Bill Grimes, and Mary M. Their families and friends deserve to know their sacrifice and contributions to protecting this nation.

We were, however, blessed in finding a superb agent in Mr. Andrew Wylie, who quickly brought us to a wonderful relationship with Skyhorse Publishing. With a contract in hand, we had only to write the book in eighteen months.

Remarkably, that process ran like clockwork. Support for the project was enthusiastic, and everyone was generous with time. Eighteen months to the day after project launch, the manuscript was completed. All that remained was a lap through the respective Publications Review Boards (PRBs) of the Air Force and CIA, to ensure that the book did not disclose classified information.

The Air Force conducted a timely, professional review of the material, thanks to some remarkable senior support from Mr. James "Snake" Clark and Mr. Ken Johns. As our story touched on defense equities outside the Air Force, the manuscript was escalated to the Office of the Secretary of Defense, Defense Office of Prepublication and Security Review. Bearing the enthusiastic support of senior Air Force leadership, who declared the book "an excellent Air Force story of innovation in the hours after 9/11" along with a gracious letter from Congressman Bacon to the Secretary of Defense, the book cleared DOD review.

The same could not be said for the review process at the CIA. From the beginning, the Agency did not want the manuscript released in any form. In fact, our first letter from the CIA's Publications Review Board declared that "the manuscript's very premise is currently and properly classified, as is its content." There is not an ounce of hype in the assertion that this is the book the CIA didn't want you to read.

Despite a legal mandate for the PRB to turn a review in thirty days, we endured almost four years of what can reasonably be described as an orchestrated campaign of stonewalling and obfuscation. At one point we were told this book could only be "released" at the highest levels of classification, and only to be read by those with the proper clearances. Then we were told to destroy it altogether.

At one point in this book, a group of Soldiers and Airmen, besieged on all sides, was saved by a Predator that flew in out of nowhere. With respect to our fight with the CIA, our Predator was Kevin Carroll, a remarkable and passionate attorney with the firm of Wiggin and Dana. Kevin and his team flew in and fired a legal volley that barely had time to clear the rails before the CIA folded its position. It was glorious and desperately needed; words do not exist to describe what Kevin's gracious help meant to all of us. It also marked the first time in the long history of this effort that we didn't have to say, "Never mind, we'll do it ourselves."

If the reader takes nothing else from this document, it should be that patriots at every level work tirelessly to challenge the system and drive meaningful change in defense of our nation and our freedoms. This book is for them, the unnamed as well as the named, the true heroes who have protected and defended the Constitution against all enemies, foreign and domestic.

Finally, and most important, thanks go to our families who have endured our ceaseless stories and rants. They all made great sacrifices, allowing us the time to write this book, after having already sacrificed so much more when we actually lived it. We are eternally grateful.

MARK'S PERSONAL COMMENTS

We've talked about this story for years, but I didn't pursue it seriously until I retired. This was never about what we did but about the great Americans that surrounded us.

I'm not sure if I was naive or stupid, but from day one I trusted Alec. Although we had our differences, we had a common bond in the belief in the importance of our task. Where most saw obstacles, we saw opportunities. I couldn't have had a better cellmate.

This book would still only be a title page had Alec not introduced Michael Marks to the project. Michael turned our rambling, disjointed, and sometimes unintelligible stories into vignettes, then chapters, and miraculously a book. I am forever grateful.

I can't overstate my gratitude to Kevin Carroll for taking on our project. Without Kevin, you wouldn't be reading these words. His legal savvy pushed the Agency where we could not. He is now an integral member of our team.

I am indebted to three senior Air Force leaders, Major General Glen Shaffer, USAF, Ret., Mr. James "Snake" Clark, and Colonel Ed Boyle, USAF, Ret., who took a chance on a young, brash major and gave me lots of responsibility and leeway but never let me hang myself. They taught me much about leadership, life, and ways to solve problems, especially in the DC Beltway. Ed continually stuck his neck out for me, for the mission, and, most important, for our team, at the risk of his career. Contrary to popular belief, let the record show he was the first armed UAV/RPA squadron commander.

Our operations team was a diverse group of officers, enlisted Airmen, civilians, and contractors. It started with Big Safari and our contractors General Atomics, L-3, Raytheon, and others rapidly meeting our challenging operational requirements at every turn. At our operational site, it wasn't about who signed the paycheck or what rank someone had; it was "one team, one fight." We came from different tribes of the Air Force: intelligence, pilots, sensor operators, communication, and weather, but any hint of tribalism was left at the gate. Some called us a ragtag bunch. We were far from perfect individuals, but in my mind, we were the perfect team for the task.

To our team, we tried our best to deconflict everyone's recollection of the events. If we erred at any point, it was not out of malice.

Some of my Air Force brethren will have preferred we use the term RPA, for Remotely Piloted Aircraft, rather than UAV or drone. But we went with the term predominantly used at the time. Rest assured, no one appreciates the man-in-the-loop essential to the success of this project more than us.

Thanks to my wife, the love of my life, Colonel Angelina Maguinness, for helping us get through the final edits. Over the years, no one has sacrificed more for my service than my three beautiful daughters, Brittany, Jaclyn, and Aidan, and they are my pride and joy.

ALEC'S PERSONAL COMMENTS

My interest in this project began in 2012 as an effort to leave some notes for my future grandkids to read someday and be as accurate as possible to the events and personalities during this pivotal period. I hold our coauthor and dear friend, Michael Marks, responsible for pulling together what is hopefully an entertaining and enlightening story from some rather dry notes. I am exceptionally grateful for his gifted writing and amazing patience.

We have stressed acknowledging the many unique personalities who were essential to the development of Predator Hellfire in this revolutionary effort. Charlie Allen, Cofer Black, Rich Blee, and Diane Killip's trust and confidence in giving me the challenge (or perhaps positioning me as a scapegoat if things were to fail) was pivotal in my career. At that point, I was too inexperienced in the ways of the Agency to know it was supposed to be hard.

My approach was simple: pull in exceptional and talented people regardless of their organization or agency and give them as much support and top cover as I could. This was made easier by having Hal M., my CIA partner in crime and occasional bodyguard as we played good cop/bad cop throughout the depth and breadth of the CIA. We knew all too well that the oddsmakers were betting against us, but who could argue with doing everything we could for the stated objective to go after the world's most wanted terrorist?

Having Mark as my counterpart was the perfect choice from the Air Force. Mark, his hand-picked team, and the Air Force leadership who selected him for the job deserve the lion's share of the credit for the technical success of the program. He and his team are among the best Americans I have ever had the privilege to work with.

The contributions of the Air Force at large and key programs like Big Safari were obviously essential. The Navy's NAVAIR 4.5x, along with Alex Lovett at OSD, turned miracles on a regular basis. The Joint Staff and unique experts like H. "Doc" Cabayan opened doors and authorities that were mission essential. DIA and Joint Staff support was critical to the foundation of our knowledge in Afghanistan. NIMA accepted the challenge and diligently worked to transition into the world of full-motion video and massive volumes of data. NSA watched our backs and cracked the vault door open ever so slightly to do so. The NSC prodded us from our comfort zone and gave us some critical tools to take the fight into Afghanistan. Importantly, our small Agency CTC team of Hal, Diane, Doc, Linda, Mary, Sharon, Brian, Al, Bob, and others set their careers and personal lives aside to champion the cause.

We all owe our families and loved ones for their extensive sacrifices of time without us during these periods of service and separation. A special thanks to my amazing family for their love and sacrifices during the extended period to write this book and prod it through the review process.

Finally, a warm welcome goes to the latest member of our extended family and our favorite attorney, Kevin Carroll from Wiggin and Dana, who broke this book loose from thirty-eight months in the PRB process.

MICHAEL'S PERSONAL COMMENTS

It means a great deal when someone trusts you with a valuable item, like tossing you the keys to the car. It means everything when a person trusts you with his or her life story. I hope the words in this book live up to the task. If there is any fault in its telling, it is mine alone.

As a writer and as an American, I consider the chance to contribute to this book a dream come true. I have been welcomed into a small circle of the most amazing people who, without exaggeration, not only changed modern warfare but also sparked the global drone revolution. As a result, we see unmanned cars and trucks on our highways and robots delivering everything from pizza to organ transplants. I believe that, in ways we cannot now even imagine, the echoes of Hellfire Predator will be felt for generations to come.

The funny thing is, you would never know it to meet this team. Diverse in so many ways, they are all endearing, quick to tell a story of how somebody else on the team saved the day or came up with a brilliant solution. They delight in each other's victories. When I began this project, both Mark and Alec made it clear that the whole reason for this book was to tell their story, the human story, not to recount a litany of technology or politics. Yes, some of this narrative stands on bones made of titanium and carbon fiber, but the beating heart inside is the men and women who worked so hard and sacrificed so much to bring it all to life. They welcomed me as one of their own, sharing fears, foibles, and triumphs with equal candor. I have seen their love for one another and felt the loss they share for comrades now departed.

I cannot recall a project I have ever done, nor presume one will ever exist, that doesn't benefit from the incredible support of Ted Deeds, Barry Solomon, Kevin Ricks, and Greg DeSantis, the latter of whom produced our breath-taking cover. You guys are without peer, and your friendship is beyond measure.

I also extend my deep thanks to Mr. Jon Arlan for his ceaseless patience and endless support as we suffered through one blown deadline after another while the CIA maintained its chokehold on this project. I could not imagine a better friend and business partner in this endeavor.

And of course, I send love to my family—to my wife, to my mom, as well as to Halina, Vera, Elizabeth, and my "Lil' Cuz" Christine. I am so proud of you all. Your love and support mean the world to me.

Writing this book, with these people, has given me one more gift: a deeper understanding of a comment made by Major Dick Winters of the 101st Airborne as told in the televised production *Band of Brothers*. In a slight paraphrase of his words, I can now look forward to the day when one of the many kids in my life will look at my wall, covered with photos and flags from strange, distant places, and ask, "Were you ever a hero, Uncle Mike?"

And I will answer, "No, but my friends are heroes."

INDEX